# Biotechnology in Sustainable & Organic Farming

## Scope and Potential

# Biotechnology in Sustainable & Organic Farming

## Scope and Potential

Vinod Kumar Jain

Oxford Book Company

Jaipur ▪ New Delhi

*Published By*

Oxford Book Company

**Regd. Off.** "Bony Residency", Gate No. 2
Opp. Tilak Public School, Vishweswariya Nagar,
Gopalpura Road, Jaipur - 302018, India
Ph: 0141 - 2761280, Telefax: 2761381
email - oxfordbookjpr@gmail.com
website - oxfordbookcompany.com

**Delhi Office:** 102, 1st Floor, Satyam House,
4327/3, Ansari Road, Darya Ganj,
New Delhi - 110002, India
Ph: 011-45652440

*Distributed By*
**KEYA INTERNATIONAL PRESS LLP**
E-mail: keyainternationalpressllp@gmail.com

Published in India, 2024



ISBN : 9788189473631

Price: ₹ 1950.00



Typeset by Shivangi Computers, Jaipur

Printed and bound in India by Thomson Press Ltd.,
New Delhi

# Preface

The cause of increasing the practice of sustainable and organic farming has found much strength in the global concern for maintaining environmental balance, reducing pollution, and increasing food quantity and quality. In such a scenario, biotechnology with its use of biological agents for industrial or manufacturing purposes such as through genetic engineering, offers a pertinent solution in conjunction with sustainable and organic farming. As the aim of such agricultural practice is to ensure the productivity as well as sustainability of agriculture, biotechnology offers the prospect of enhancing this goal and making it more achievable.

The present text aims to delineate for the readers the role of biotechnology in enhancing sustainable and organic farming. It analyses and makes assessment of the scope and potential of the field especially as the use of both sustainable and organic farming increases and biotechnology itself advances to new revels. In addition the text incrorporates current trends and developments in the field, especially those which are pertinent to sustainable organic farming. Critical perspectives have been included too which look into the future prospects offered.

*Binod Kumar Jain*

## Contents

# 1

# Introduction

Biotechnology provided very high profit investment opportunities in the developed countries during the last decade. The applications of these technologies were extended to a wider area to improve the services, while taking advantage to expand the business opportunities by the inventors. These opportunities for biotechnology could not directly benefit the developing countries in the same magnitude, because of patent rights and lack of infrastructure to extend the benefits to the society. As most of the advanced inventions were carried out with financial support from private sources, the developed countries could not have direct access to the research findings. Nevertheless, series of research projects on biotechnology were sponsored by the Ministry of Science and Technology in India. It may need huge investment and long time to make any significant break in this field. Even in the absence of any major breakthrough, biotechnology can provide good scope for setting up small scale business enterprises, particularly in the agricultural sector.

India is one of the leading developing countries making economic progress through industrial and agricultural development programmes. Agriculture is the major source of livelihood, particularly in rural areas, where about 65-70% of the people have been living. However, the present level of agricultural production has not reached the optimal stage because of series of hurdles. Major bottlenecks among them, are lack of resources such as water nutrient and good quality planting material, improper management of pests and diseases and poor harvest management of the produce.

As against the world average of 172 kg/ha chemical fertilisers, Indian agriculture consumes only about 70 kg/ha. The average fertiliser consumption would be even lower, if the fertiliser applied for three important crops like paddy, wheat and sugarcane were kept out of the average. Thus, a majority of the crops suffer due to nutritional deficiency. In the areas of intensive agriculture such as Punjab, Hariyana, Tarai region of Uttar Pradesh and the sugarcane belts in the rest of the country, excessive doses of imbalanced fertiliser application has been a problem of serious concern. These areas also suffer from excessive use of water for irrigation, when soils turn alkaline or saline and the fertilisers applied to the soils are not available for the crops.

Biotechnology can help to solve these problems in two ways. Firstly, for crops where fertiliser application is very low, bio-fertilisers can fix atmospheric nitrogen and provide micro-nutrients useful to plant growth. Use of blue-green algae has also been beneficial to rice crop. Microbes such as mycorrhizae have been helpful to overcome the stress from drought and diseases. Increase in the cost of fertilisers even in Western countries has forced their farmers to use bio-fertilisers for many crops.

## HIGH YIELDING CULTIVATION

Intensification of agriculture through massive adoption of high yielding cultivation, increased use of synthetic inputs like chemical fertilisers and pesticides, greater exploitation of irrigation potentiality of surface and ground water resources and farm mechanisation have largely been responsible for a spectacular achievement in the food grain production that we have achieved over past three decades.

Paradoxically however overexploitation of natural and renewable resources and indiscriminate and irrational use of synthetic inputs like inorganic fertilisers and pesticides in view of producing more and more from unit piece of land are being increasingly realised to seriously impair the ecological balance and putting the environment in jeopardy.

High yielding cultivation is more fertiliser responsive which often led to aggravation of pest problem as the plants become succulent enough to be fed upon by a variety of crop pests. This in turn necessitates increasingly huge amount of pesticides to combat pest problem. Increased use of pesticides has emerged as a potential source of danger to sustainability of environment that endangers the existence of all forms of life on this planet. Perils and pitfalls of pesticides have been well evidenced due to their residual toxicity in our food chain.

From a number of trials conducted across the country, toxic residues of pesticides have been revealed in the food stuff not only of plant but also of animal origin like milk and milk products, fish, meat and egg etc. at concentrations much higher than the permissible level of human body. Therefore the apparent contradiction of our necessity for nutritional security on one hand and environmental sustainability on the other makes it inevitable to resort to the organic or eco-farming system as it appears to be a possible option to meet both these objectives.

The later implies a farming system that primarily aims at cultivating land and raising crops under ecologically favourable condition. It emphasizes restricting the use of chemical inputs whether it is inorganic fertilizers or pesticides and instead, relies more on an integrated approach of crop management practices making use of cultural, biological and natural inputs. Addition of organic manures such as FYM, recycling of organic wastes through composting, green manures and biological inputs like vermicomposts and biofertilisers etc. constitute important components for plant nutrient management in organic farming. Similarly it also takes utmost advantage of the natural mechanism for pest management with utilisation of bioagents such as predators and parasites available in nature in plenty and the botanical pesticides which are effective in controlling crop pests posing no risk to the environment. Agronomic practices such as crop rotation with judicious selection of crops, inter cropping and companion cropping, stubble mulching and use of resistant varieties are among the important factors contributing to organic farming.

## ORGANIC MANURES

Organic manure in a broad sense includes composts from rural and urban wastes, crop residues, agro industrial bio wastes and green manures, apart from the commonly used FYM. Organic manure improves soil physical condition including soil porosity and water holding capacity and microbial environment, replenishes essential micronutrients in soil, increases the utilisation efficiency of applied fertilisers and favours micronutrient availability to the plant. Organic manure is of paramount importance not only in augmenting the crop production but also for making the agriculture sustainable as an ecofriendly means of soil health management. It is well established that FYM plays an additional role than its capacity to contibute NPK. Unlike chemical fertilisers that supply only the major nutrients, FYM is a store house of several plant nutrients and acts as a good soil conditioner.

Green manuring has recently been under practice by our farmers for decades. Estimates suggest that a 40-50 days old green manure crop can supply up to 80-100 kg. N/ha. Even if half of this N is crop utilisable, a green manure crop can be a substitute to 50-60 kg. fertiliser N/ha. Some of the potential green manuring legumes are dhanicha, sunhemp, cowpea, mung, bean, guar and berseem etc. Dhanicha, sunhemp, mung bean and guar grown during kharif season as green manure crops have been reported to contribute 8-21 tones of green matter and 42-95 kg. of N/ha. Similarly, Khesari, cowpea and berseem grown during rabi season can contribute 12-29 tons of green matter and 67-68 kg of N/ha.

## VERMICOMPOSTING

Vermicomposting is an effective means of composting the decomposable organic wastes using earthworms naturally present in the soil. Vermicomposting is a mixture of worm casts enriched with macro and micronutrients (N, P, K, Mn., Fe, Mo, B, Cu and Zn.), some growth regulating substances such as gibberellins and auxins) and useful micro flora (Azospirillum, Actinomyces and Phosphobacillus) etc. The nutrient level of vermicompost (1-1.5%N, 0.6-0.8% P and 1.2-1.5 K) is higher than any other compost. From the available information it is well documented that earthworms can consume all types of organic matter and convert them into available form of nutrients. Vermicompost improves the physical and biological condition of soil, improves soil fertility and pulverises it through their churning and turning action in addition to contributing plant nutrients, improves aeration and water holding capacity. It is reported that soils with casts contain 5 times nitrogen, 7 times phosphorus, 11 times potash, 2 times magnesium and 7-8 times actinomyces more than in soils without earth worm casts. It being a natural means of soil fertility management fits well into integrated plant nutrient management strategy for sustainable agriculture.

## MECHANISM OF BIOLOGICAL NITROGEN FIXATION

Atmosphere containing as much as 78% nitrogen can be a potential source of this essential nutrient to soil utilising the nature's own mechanism of biological nitrogen fixation. It not only offers an economic and ecofriendly source of nutrient supply, moreover evidences show that nitrogen so derived is less prone to loss than fertiliser nitrogen and as a long term effect it builds up a reserve of readily mineralisable organic nitrogen. Thus it plays an important role in sustainable agriculture, reduces the requirement of externally applied nitrogen fertiliser and favourably influences the soil microenvironment.

BNF is a natural system of biological mobilisation of atmospheric nitrogen which can be easily available to and utilised by plants mediated by microorganisms like some eubacteria, cynobacteria, actinomyces and a few archaebacteria which are commonly called diazotrophs. They have the ability to convert atmospheric elemental nitrogen ($N_2$) to ammonia ($NH_3$) with the help of nitrogenase enzyme system. The efficacy of true bacteria like Rhizobium, Azotobacter, Azospirillum and cynobacteria like Anabaena, Aulosira, Nostoc, Plectonema and Tolypothrix etc. have already been well established under field conditions and such microbial inoculants have gained wide acceptance due to their high nitrogen fixing potentiality.

**Minimum Tillage**

The conventional tillage operation has disastrous effect on soil erosion causing greater loss of nutrients. This has led to necessity of stubble mulching. In the stubble mulching the soil is protected by the crop residues left on the soil surface during fallow periods. This has now become well established that in stead of frequent tillage operations minimum tillage is more useful not only because it offers cost effectiveness but also contributes to conservation of soil and moisture. Available information reveals that it gives as good as or even better yield than the conventional tillage practices. It aims at reducing tillage to the minimum, necessary for ensuring a good seed bed, rapid germination, satisfactory crop stand and favourable growing conditions.

### Selection of Crops for Crop Rotation, Inter-cropping and Companion Cropping

Crop rotation has an important role to play in organic farming. Judicious selection of crops in crop rotation helps in efficient utilisation of plant nutrient from different depths of soil. The practice of inter cropping also helps in minimising the risk of crop failure due to uncertain rainfall and infestation of pest and diseases. Available information reveals that in case of inter cropping of sorphum and red gram at 2:1 row ratio it resulted in 70% higher yield over the individual cropping of both the crops. Besides it also reduced the wilt diseases of red gram. Another advantage was that nitrogen fixed by the red gram could be utilised by the sorghum crop. Another trial of companion cropping comprising of three rows of gram followed by one row of linseed revealed that incidence of pod borer in gram as well as fly in linseed was remarkably reduced and yield also increased by about 14% in this inter cropping over the sole cropping of both the crops.

### Integrated Pest Management

The multifarious harmful consequences of indiscriminate use of pesticides have cropped up as a serious threat to the ecosystem. In view of the fact that increased use of pesticides has been drastically endangering the environmental sustainability, integrated approach to pest management needs adequate importance to make the agriculture ecofriendly. Like integrated plant nutrient management, IPM also makes use of an ideal combination of physical, chemical, biological and cultural methods to contain pest damage with minimum ecological implication. It takes the utmost advantage of natural mechanism of pest suppression. Modifications of crop environment to make it unfavourable for infestation of pests, use of resistant cultivars and transgenic alongwith a judicious and need based use of preferably safer chemicals are the major components of IPM. Biocontrol is an integral component of integrated pest management strategy. It employs natural mechanism of pest suppression in which the natural agents like predators, parasites and microbial agents suppress the pest population. If the natural enemies can be effectively conserved, the need of other control will automatically reduce.

**Rhizobium Azotobacter**

Use of biofertilisers carrying Rhizobium and Azotobacter have been in practice since a long time. However, the technology did not catch up till recently probably because of inadequate awareness, lack of marketing and insignificant results in the field. These bio-fertilisers were generally distributed free of cost by the Agricultural department mostly to poor farmers and neither the promoters nor the users were serious about the use or benefits. This product came to the field at the last moment without any planning and hence farmers would not have taken this as a serious input for their crops.

Biofertilisers produced mainly by the government sponsored laboratories, had no marketing expertise to launch the product in the field. Hence, biofertilisers could not establish their superiority before a well planned marketing to launch the product in the field. The third and the most prominent cause of poor uptake was poor quality product. Most of the biofertiliser laboratories did not have adequate production and quality control facilities and hence there have been wide variation. The carrier used was not of a standard quality.

After the acidity nature, the carrier killed the bacteria before reaching the field. These problems have now been solved to a great extent through the application of biotechnology combined with a marketing strategy to organise the sale through the fertiliser marketing network without any cut throat competition. The technologies have been further simplified to use the culture in liquid form, which can have a long shelf life and are less expensive.

Further, the focus for promoting biofertiliser has shifted from more intensive crops like groundnut, wheat paddy, sugarcane to other crops where farmers have not been applying adequate doses of fertilisers. This has helped to improve the crop yields while reducing the cost of production. Those who have realised the benefits are now coming forward to purchase biofertilisers on their own without any subsidy from the government. This is a good trend which ensures the growth of this sector. Soils having high Phosphorus have the problem of availability of phosphorus to plants and phosphate solubilising micro-organisms (PSM) can be useful to reverse this process.

Vermicomposting is also an extension of biotechnology, which is being reinvented with the growing awareness on organic farming. Earthworms constitute more than 80% of the soil invertebrate biomass. They play a significant role for maintaining the soil productivity. Earthworms enhance the decomposition of organic matter and they also contribute 20-100 kg nitrogen per ha per year, besides other mineralised nutrients and plant growth factors. There is a need for promoting grass-root level vermiculture promotion units to supply superior quality culture to farmers for production of vermicompost.

## BIOTECHNOLOGY AND PLANT PROTECTION

Application of biotechnology for plant protection has been gaining popularity during recent years.

### Biotechnology and Microlevel Entrepreneurship

These are some areas where technologies have been preferred for commercial production. However, the success lies in taking these technologies to the rural people at an affordable cost. As these technologies are not able to provide very high margin, due to poor purchasing power, larger business houses are not keen to take up these ventures. Thus it can be a blessing for promoting small-scale village level enterprises which is a network to receive technical information and marketing support.

The techniques of bio-fertiliser and bio-pesticides production needs a small investment to install fermentors with accessories for packing and storage. The critical support needed is good quality germplasm and control on the quality of the fertiliser. In case of verimiculture, the technique is still very premature and further development is needed for faster regeneration of worms. Present scale of investment is very low and the gestation period is also very short

For ensuring success of small scale bio-technology enterprises, the scheme can be effectively operated on a franchise. A central level organisation having facilities for providing equipment, culture and technical support can give licence to individuals to operate the centre with a brand name or trademark. The central agency apart from providing inputs can also take up product promotion, awareness and field monitoring. They can also coordinate distribution of the produce to bridge the surplus and shortages in different localities. Such a coordination network will surely extend the benefits of biotechnology to even the poorest farmers in the country.

**REFERENCES**

Lampkin, L., *Organic Farming,* Farming Press, UK, 1990.

Lampkin, N & Measures, M., *Organic Farm Management Handbook,* University of Wales, Aberystwyth, 1999.

Natarajan, M & Willey, R.W., "Sorghum-pigeonpea intercropping and the effects of plant population density" *Journal of Agricultural Science,* 1990.

# 2

# Traditional Organic Farming Practices

The indiscriminate use of chemical fertilisers, hybrid seeds and pesticides has resulted in various environmental and health hazards coupled with socio-economic problems. Though agricultural production has continued to increase, the rate of yield per hectare has started to decline. The causes of the environmental crisis are in fact rooted in the prevalent materialistic paradigm, which promotes high input technologies and practices in all sectors (domestic, agriculture, industrial, service) that lead to soil erosion, salinisation, all types of pollution, desertification and loss of biodiversity.

In the agriculture sector, the green revolution selected crops for high yield and palatability. By sacrificing natural resistance for productivity, it made them more susceptible to pests. Since the structural base of agricultural systems is maintained as a monoculture, pest problems will continue on a negative treadmill that reinforces itself, as more and more vulnerable crops call for increasingly destructive or expensive high-tech protective measures.

The fruits of the green revolution are turning sour due to the stagnation in crop productivity combined with deterioration in the quality of food and soil health. These developments have led to doubts about the long-term consequences of existing practices. Thus the gain in production is very much associated with pain of different kinds and magnitudes.

The results of the green revolution have proved to be a paradox: on the one hand, the green revolution offered technology as a substitute both for nature and as well as politics, by the creation of abundance and peace. On the other hand, the technology itself demanded more intensive natural resource use along with intensive external inputs and a restructuring of the way power was distributed in society. While treating nature and politics as dispensable elements in agricultural transformation, the green revolution created major changes in natural ecosystems and agrarian structures.

Sir Albert Howard, who was associated with the Pusa Agriculture Research Centre, made an almost prophetic declaration, right at the beginning of the twentieth century about the emerging practices of modern farming. He said: "These mushroom ideas of agriculture are failing, mother earth deprived of her manurial rights is in revolt; the land is going on strike; the fertility of the soil is declining......Soil is no longer able to stand the strain. Soil fertility is rapidly diminishing particularly in the US, Canada, Africa, Australia, New Zealand. The loss of fertility all over the world is indicated by the growing menace of soil erosion..... Diseases are on the increase... the diseases of crops and animals which feed on them". Though they may have sounded like an exaggeration at that time, his predictions have all come true in a magnified way.

As history shows, former civilisations were able to overcome their economic and cultural declines when the ecosystems which made up their environment remained intact and free of interference. Avoiding detrimental changes in the material cycles and energy fluxes and preventing the loss of biological diversity in our natural environment are of utmost priority among the goals of sustainable development.

Soil is the basis of all human life. Destruction of the soil has contributed to the fall of past civilisations yet the lessons of history are seldom acknowledged and usually unheeded. The only hope for a healthy world rests on re-establishing the harmony in the soil that has been disrupted by the modern methods of chemical farming and unplanned rapid industrial growth. These methods cause serious problems to land degradation. Today's cropland losses impair the well being of the living as well as of the generations to come.

## DEVELOPMENT OF SUSTAINABLE AGRICULTURE

The idea of sustainable agriculture is a response to the decline in quality of the produce and of the resource base associated with modern farming. It captures a set of concerns about agriculture conceived as the result of the co-evolution of socio-economic and natural systems. Agricultural development resulting from a complex interaction of a multitude of factors, and a wider understanding of the agricultural context requires a study of the relations between farming, the environment and social systems. It is through this deeper understanding of the ecology of farming that doors will open to new technological and management options more in tune with the aims of a truly sustainable agriculture.

The goal is to develop agro-ecosystems with minimal dependence on high agrochemical and energy inputs and in which ecological interactions and synergy between biological components provide the mechanisms for the systems to sponsor their own soil fertility, productivity and crop protection. The five objectives of productivity, security, protection, viability and acceptability are called the five pillars of sustainable land management and they must be achieved simultaneously if true sustainability is to be predicted. Attainment of only one or several of the objectives, but not all, will result in partial or conditional sustainability.

In fact, sustainability is not possible without preserving the cultural diversity that nurtures local agriculture. A closer look at ethno-science (the knowledge system of an ethnic group that has originated locally and naturally) will indicate that local people's knowledge about the environment, vegetation, animals and soils is enormous. Peasant knowledge about ecosystems usually results in multi-dimensional land use production strategies, which generate, within certain ecological and technical limits, the food self-sufficiency of communities in particular regions.

Stable production can only take place within the context of a social organisation that protects the integrity of natural resources and nurtures the harmonious interaction among humans, the agroecosystem and the environment. Sustainable development is development that meets the needs of the present without compromising the ability of future generations to meet their own needs. The basic tenets of a sustainable agriculture system are the conservation of renewable resources, adaptation of the crop to the environment and maintenance of a moderate but sustainable level of productivity. And it should be economically viable and socially acceptable.

The production system must:

— reduce energy and resource use and regulate the overall energy input, so that the output: input ratio is high;
— reduce plant nutrient losses by effectively controlling leaching, runoff and erosion, and improve nutrient recycling through the promotion of legumes, organic manure and compost and other effective recycling mechanisms viz., residue management;
— encourage local production of feed items adapted to the natural and socio-economic setting;
— sustain desired net output by preserving the natural resources (by minimising soil degradation);

— reduce costs and increase the efficiency and economic viability of small- and medium-sized farms, thereby promoting a diverse, potentially resilient agricultural system.

From a management viewpoint, the basic components of a sustainable agro-ecosystem include:

— vegetative cover as an effective soil- and water-conserving measure, met through the use of no-till practices, mulch farming, use of cover crops, etc;

— a regular supply of organic matter through the regular addition of manure and compost, and promotion of soil biotic activity;

— nutrient recycling mechanisms through the use of crop rotations, crop/livestock systems, use of legumes, etc.

— pest regulation assured through enhanced activity of biological control agents, achieved by conserving and multiplying natural enemies in an eco-friendly way.

The ultimate goal of sustainable agriculture system design is to integrate farm components in a holistic fabric so that overall biological efficiency is improved, biodiversity is preserved, and agroecosystem productivity and its self-regulating capacity are maintained. The idea is to design an agro-ecosystem that mimics the structure and function of local natural ecosystems, that is, a system with high species diversity and biologically active and conserved soil, one which promotes recycling and prevents resource losses. A major strategy in sustainable agriculture is to restore agricultural diversity in a given time and space through alternative cropping systems, such as crop rotations, cover crops, inter-cropping, border cropping or crop/livestock mixtures all of which exhibit several ecological features.

### Emerging Trends in Sustainable Agriculture

Modern agriculture is not sustainable as it is not in consonance with economics, ecology, equity, energy and the socio-cultural dimension. The world over, therefore, environmental degradation sourced to agriculture is reaching catastrophic proportions. Time is running out if we are going to continue with more or less the same strategies in agriculture viz., high agro-chemical responsive hybrids, monoculture, ex-situ biocontrol agents, terminator seed and like technologies. Ironically, the shift from chemical farming to sustainable agriculture is being sought within a very narrow vision, posing very severe irreparable risks to the ecosystem in the long run.

An alternative sustainable farming system, which is ecologically sound, economically feasible and socially just, is the need of the hour. Sustainable agriculture is a unifying concept, which considers ecological, environmental, philosophical, ethical and social impacts balanced with cost effectiveness.

Several aspects of traditional systems are relevant, such as their knowledge of farming practices and the physical environment, biological folk taxonomic systems, or use of low-input technologies. By understanding the ecological features of traditional agriculture, such as the ability to bear risk, production efficiencies of symbiotic crop mixtures, recycling of materials, reliance on local resources and germplasm, exploitation of the full range of micro-environments, etc., it is possible to obtain important information that may be used for developing appropriate agricultural strategies tailored to the needs, preferences and resource bases of specific peasant groups and regional agro-ecosystems.

Among the various alternatives, organic farming is gaining acceptance through out the globe as it has the potential to provide practical solutions to mitigate the maladies afflicting conventional or modern farming. Before India faced the onslaught of chemical farming, its thinkers like Vinoba, Gandhiji, Kumarappa could visualise the future of Indian farming through their non-violent approach – aptly suited to the present.

Indian culture, nurtured in the lap of nature, in the forests, and in the verdant valleys of rivers coursing down from the mighty Himalayas, imbibed a deep sense of oneness with all things natural. Ancient (Vedic) culture taught the veneration of the earth as mother, the sky as father, the air as prana (soul), the sun as energy and water streams as life sustaining veins.

### Alternative Agriculture Systems

The goal of an alternative agriculture system is to enable the peasants to become the architects and actors of their own development. From a management perspective, the objective of such system is to provide a balanced environment, sustained yields, biologically mediated soil fertility and natural pest regulation through the design of diversified alternative agricultural systems and the use of low-input technologies. The strategy is generally based on ecological principles so that management aims at optimal recycling of nutrients, organic matter turnover, closed energy flows, water and soil conservation and balanced pest/natural enemy populations. The idea is to exploit the complementarities and synergisms that result from various combinations of crops, trees and animals. The optimal behaviour of an alternative agricultural system depends on the level of interaction between its biotic and abiotic components. By assembling a functional biodiversity, it is possible to provoke synergisms. These in effect subsidise alternative agriculture processes, by providing ecological services such as the activation of soil biology, the recycling of nutrients, or the enhancement of beneficial arthropods.

Today there is a whole battery of practices and technologies available that vary in effectiveness as well as in strategic values. Some, which include practices that are already part of conventional farming are of prophylactic value, while others which are key, are of a preventative nature and act by reinforcing the 'immunity' of the agro-ecosystem. The effects of many of these practices have been scientifically documented and tend to have wide geographic implications.

These technologies do not emphasize the boosting of yields under optimal conditions, as green revolution technologies do; rather, they assure consistency of production under a whole range of soil and climatic conditions – especially the marginal conditions that usually prevail in small-farm agriculture.

The need, however, is not to focus on particular technologies, but rather on an agroecosystem management approach which emphasizes crop diversity, use of manures, green manures, urban and rural wastes, legumes in rotations, animal integration, recycling and use of biomass and residue, and incorporates an assemblage of suitable alternative technologies. Further, those kinds of seed which provide food, fodder, soil fertility and diversity should be given due priority in such systems.

The role of an alternative agricultural system should not be limited to input substitution alone, but to ensure that it is economically and ecologically sustainable. Alternative agriculture should be successful in addressing the ecological causes of environmental problems in agriculture which are in fact deeply rooted in the monoculture structure prevalent in large-scale agro-production systems.

Various alternative agricultural systems have been developed globally. These include traditional and natural farming, organic agriculture, ecological farming, vedic agriculture, permaculture, biodynamic farming and LEISA (Low external input sustainable agriculture). The real success of these systems on an evolutionary time scale demands that the question of the paradigm of development and the technology package should be considered together.

## SUSTAINABLE AGRICULTURAL SYSTEMS

### Vedic Agriculture

Vedic agriculture pertains to the knowledge of farming that is presented in the Vedas, the age-old Indian scriptures originating 10,000 years ago. This system of farming makes use of various forms of inputs and not just some preparations made for providing nutrition or managing problems. The Vedic way of agriculture comprises: (a) farming operations and (b) *yajnas* and *ishtis*.

(a) Farming operations include biological practices of cultivation. The unique feature in the Vedic system is performing the cultivation practices at a specific (right) time. There is a lot of significance invested in the panchamahabhutas i.e., the five basic elements: prithvi (earth), aap (water), tejas (fire or light), vayu (air) and akasha (space). Bhutas means living creatures. Mahabhutas are prior and superior to living beings since the very existence of the latter is totally dependent on them. The mahabhutas are the primary natural resources essential for all agriculture. These panchamahabhutas have a significant effect on all living beings including plants. The time bound nature of the activities helps to harness the maximum positive effect of the panchamahabhutas. This can be practically achieved by synchronising all the activities as per the Panchang or planting calendar. The planting calendar is a very important, no-cost input that can be easily adopted.

Another principal emphasis in vedic agriculture is on the importance of the cow. It is said in our scriptures that all the deities have their residence in different portions of the cow's body. A variety of preparations can be made from cow dung, cow urine, cow ghee, cow milk and curd made out of cow's milk.

The preparations which are commonly made from these sources and that are widely used as inputs in agriculture are:

— Panchagavya
— Amrit pani
— Matka khad

In addition, a variety of preparations for nutrition and pest and disease management can be made available by utilising various medicinal herbs and other plants. Medicinal herbs like shatavari (*Asparagus* spp.), *bhallataka* (marking nut), *ashwagandha* (winter cherry), *kana* (long pepper), *guggulu* (*Comniphora mukul*), asana (*Terminalia tomentosa*), *shatapushpa* (dill seeds), *triphala, karanja* (*Pongamia glabra*), *adathoda*, etc., and other plants like *haridra* (turmeric), ginger, oleander, karavira (*Nerium*), *tulasi* (*Ocimmum spp.*), tobacco, *langli* (*Gloriosa superba*), arka (*Calotropis*) and *sami* (*Prosopis spp.*) are also used in plant protection.

(b) *Yajnas* or *ishtis* are the other major component of vedic farming. The yajna is an offering of specific ingredients offered to a fire prepared from specific materials in a pyramid pot of copper or earth or altars of specific shapes at stipulated timings. *Yaga, homa, ishtis* and *agnihotra* are different names or forms of yajna. A yajna is performed using various procedures and therefore is known by different names. Yajna, homa and ishtis are performed on specific occasions. The special feature of homa farming lies in the generation and utilisation of the subtle energy released by agnihotra for the activation of organic manure and bio-energy. It is based on the principle that you heal the atmosphere and the healed atmosphere heals you. It replenishes the nutrients that pollution robs from the environment.

### Organic Agriculture

Organic agriculture as widely known also involves the basic cultural practices of farming and the use of certain inputs for providing nutrition and protection to the plants. The USDA defines organic farming as a system that is designed to produce agricultural products by the use of methods and substances that maintain the integrity of organic agricultural products until they reach the consumer. The basic practices include crop rotation, mixed cropping, intercropping, strip cropping, trap cropping, mulching, tillage operations, use of natural predators, etc. These practices help to manage the problems of pests to a large extent. In addition, the inputs used in organic farming are:

***Vermicompost***

Vermicompost serves a dual benefit: it provides good quality manure with natural nutrients for the soil and plants; it can also solve the problem of farm wastes.

***Biofertilisers***

Biofertilisers have a tremendous potential for providing nutrients to the soil and plants by converting these nutrients from a non-available to an available form by the biological process of nitrogen fixation and phosphate solubilisation, etc. The microorganisms widely used, singly and in combinations, for these purposes are azotobacter, azospirillum and blue green algae, azolla, etc. All can meet the nitrogen requirements of the crops to a considerable extent. However, there are many other microbes which mobilise nutrients and improve the quality of soils in different ways. Their efficiencies should be tapped under sustainable agriculture.

***Botanicals***

Botanicals are preparations made from specific plants for the purpose of managing pest and disease problems faced in agriculture. A wide and varied base of such plants exists whose different parts can be utilised in making botanicals. A few of these can be immediately listed: Adathoda zeylanica, Adathoda vasica, Agave americana, Allium sativum, Anacardium occidentale, Argemone mexicana, Azadirachta indica, Brassica campestris, Capsicum frutescens, Cassia tora, Cinnamomum verum, Ocimum sanctum, Parthenium hysterophorus, Sesamum indicum, Annona squamosa, Melia azadiract, Hardwikia binata, etc.

*Biodynamic agriculture*

Biodynamic agriculture is a holistic system of farming developed by Dr. Rudolf Steiner in 1924. This system of farming is similar in many fundamentals to the traditional systems being practised in India since ages. Biodynamic farming is based on a systematic and synergistic harnessing of energies from the cosmos, earth, plant and cow. It includes certain dynamic practices in addition to the basic biological practices of cultivation. The inputs principally utilised within this system are:

a) Biodynamic (BD) preparations numbering from 500–508;
b) Liquid manures;
c) Cow pat pit manure (CPP);
d) Biodynamic stem paste.

Biodynamic preparations numbered 500 and 501 are made by using cow dung and quartz silica. The other seven are made by fermenting parts of certain herbs in a prescribed manner. These herbs include yarrow, chamomile, stinging nettle, oak bark, dandelion, valerian and equisetium. These preparations are called compost preparations and are utilised for making the biodynamic compost. They are meant to activate a synthesis of different nutrients in the final biodynamic compost. These are components of biological agriculture and provide it (and the ecosystem) with long-term sustainability. Other preparations like liquid manures, CPP and biodynamic stem paste are utilised for a number of cultural practices in farming.

Another unique feature of biodynamic agriculture is the use of a biodynamic planting calendar for various farming operations. Efforts are made to enrich the life of the soil by restoring its fertility in the form of humus, and by skilful application of appropriate crop rotations.

### *Permaculture*

The word 'permaculture' was coined in 1978 by Bill Mollison, an Australian ecologist, and David Holmgren, one of his students. It is an amalgamation of the ideas of 'permanent agriculture' and 'permanent culture'.

Permaculture is about designing ecological human habitats and food production systems. It is a land use and community building movement that strives for the harmonious integration of human dwellings, microclimate, annual and perennial plants, animals, soils, and water into stable, productive communities. The focus is not on these elements themselves, but rather on the relationships created among them by the way we place them in the landscape. This synergy is further enhanced by mimicking patterns already found in nature.

The central theme of permaculture is the design of ecological landscapes that produce food. Emphasis is placed on multi-purpose plants, cultural practices such as sheet mulching and trellising, and the integration of animals to recycle nutrients and graze weeds.

However, permaculture entails much more than just food production. Waste water treatment, recycling, and land stewardship in general are other important components. More recently, permaculture has expanded its purview to include economic and social structures that support the evolution and development of more permanent communities, such as co-housing projects and eco-villages. As such, permaculture design concepts are applicable to urban as well as rural settings, and are appropriate for single households as well as for whole farms and villages.

Permaculture principles focus on thoughtful designs for small-scale intensive systems that are labour efficient and which use biological resources in place of fossil fuels. Designs stress ecological connections and closed energy and material loops. The core of permaculture is design and the working relationships and connections between all things. Each component in a system performs multiple functions, and each function is supported by many elements. Many of the appropriate technologies advocated by permaculturists are well known. Among these are solar and wind power, composting toilets, solar greenhouses, energy efficient housing and solar food cooking and drying.

Farming systems and techniques commonly associated with permaculture include agro-forestry, contour plantings, hedgerows and windbreaks, and integrated farming systems such as pond-dyke aquaculture, aquaponics, intercropping and polyculture. Gardening and recycling methods common to permaculture include edible landscaping, companion planting, trellising, sheet mulching, solar greenhouses, spiral herb gardens, and vermicomposting. Water collection, management, and re-use systems like rain catchments, constructed wetlands, aquaponics (the integration of hydroponics with recirculating aquaculture) and solar aquatic ponds also play an important role in permaculture designs.

### Effective Microorganisms (EM)

For many years, soil microbiologists and microbial ecologists have tended to differentiate soil microorganisms as either beneficial or harmful according to their functions, particularly how they affect soil quality, plant growth and yield and plant health. Beneficial microorganisms are those that can fix atmospheric nitrogen, decompose organic wastes and residues, detoxify pesticides, suppress plant diseases and soil-borne pathogens, enhance nutrient cycling, and produce bioactive compounds such as vitamins, hormones and enzymes that stimulate plant growth. Harmful microorganisms are those that can induce plant diseases, stimulate soil-borne pathogens, immobilise nutrients, and produce toxic and putrescent substances that adversely affect plant growth and health. A more specific classification of beneficial microorganisms has been suggested by Professor Terua Higa, University of the Ryukyus, Okinawa, Japan which he refers to as 'Effective Microorganisms' or simply, EM.

The consortium of effective microorganisms (EM) was developed by Professor Higa in his laboratory. EM consists of mixed cultures of beneficial and naturallyoccurring microorganisms including predominant populations of lactic acid bacteria and yeasts and smaller numbers of photosynthetic bacteria, actinomycetes and other types of organisms all of which are mutually compatible with one another and can coexist in liquid culture.

EM can be directly applied as an inoculant to increase the microbial diversity of soils and plant. Research has shown that the inoculation of EM cultures to the soil/plant ecosystem improves soil quality, soil health and the growth, yield, and quality of crops. EM is not a substitute for other management practices. It is, however, an added dimension for optimising our best soil and crop management practices including crop rotations, use of organic amendments, conservation tillage, crop residue recycling, and biocontrol of pests. If used properly, EM can significantly enhance the beneficial effects of these practices.

#### Natural Farming

The concept of 'natural farming' was first developed by Masanobu Fukuoka, a Japanese ecologist. Natural farming advocates a 'donothing' approach to agriculture. Fukuoka's approach has similarities to organic agriculture and permaculture, but also has many features that are unique such as not preparing compost, and using seed balls instead of sowing or planting seedlings. The four basic tenets of natural farming are:

a) No cultivation – that is, no ploughing or turning of the soil: the earth should be left to cultivate itself naturally by the penetration of plant roots and the activity of micro-organisms and earthworms.
b) No chemical fertilisers – left alone the soil can maintain its own fertility through the cycle of plant and animal life. The only nutrient used is a light layer of poultry manure on top of a straw and clover mulch and this is usually provided by ducks and chickens roaming free.
c) No weeding by tillage or herbicides – weeds play a part in building soil fertility and should be controlled by mulch and other natural means rather than eliminated.
d) No pesticides – the best approach to disease and insect control is to grow sturdy crops in a healthy environment. Natural predators should be encouraged to take care of any unwanted pests.

These different sustainable agricultural systems provide us with an opportunity to transform modern agriculture into an ecologically sound, socially just and economically feasible activity. It is within the context of what has been related above that the present report attempts to assess the current status of inputs available for organic farming in India.

## TRADITIONAL PRACTICES IN AGRICULTURE

Tradition is an opinion, a belief, a custom or knowledge handed down from ancestors to posterity. In other words, it refers to the unwritten laws of ancient wisdom propagated by word of mouth which are replicable. The traditional wisdom relating to agriculture dates back to about 12,000 years when the first plants were domesticated by human beings. This wisdom has been evolving since then through accumulating experience in dealing with the situations and problems and has been recorded and channeled to the generations down the line, which can be very well found in our ancient literature.

Our ancient literature – which was most likely composed in between 6000 BC and 1000 AD – contains complete information on agriculture. This literature includes the four Vedas, the nine Brahmanas, the Aranyakas, the Sutra literature, the Sushruta Samhita, the Charaka Samhita, the Upanishads, the epics Ramayana and Mahabharata, the eighteen Puranas and texts such as the Krishi-Parasharas, Kautilya's Arthashastra, the Manusmriti, Varahamihira's Brhat Samhita, the Amarkosha, the Kashyapiya-Krishisukti and Surapala's Vrikshayurveda. Kautilya's Arthashastra deals with the agriculture of his time. Vrikshayurveda provides the complete knowledge on how to combat the problems of the plants through various traditional practices utilising available resources. One finds even in the poems of the poet Ghagh (kahawaten), descriptions about agro-managements, the timing and forecasting of weather, and crop yields.

Traditional farming systems appear to be quite complex and advanced as they exhibit important elements of sustainability: for instance, they are well adapted to the particular environment, rely on local resources, are decentralised and tend to conserve the natural resource base overall. The ancient texts referred to contain complete information about farm implements to be used, types of lands, forecast of monsoon rains, manures, irrigation, seed and sowing, pests and their management, horticulture, etc. The fertile status of the soils in most parts of our country is a result of the wisdom of our forefathers.

Farmers work with highly dynamic knowledge systems which coevolve with the time as circumstances change.

They learn and unlearn, choosing the appropriate knowledge in their struggle for livelihood. While scientists rely on averages, the knowledge of the local people is more dynamic and up-to-date, continually revised as conditions change. The integration of scientific knowledge systems with indigenous knowledge systems is vital to make agriculture sustainable. We need to maintain the soil's health in the interest of future generations. 'Some civilisations view the soil as sacred, inviolate, something which must be handed down to the coming generations intact, if not improved.' Even today, it has proven to be difficult to find terms that are the equivalent of ethnic names of soils in many classification systems. Thus the integration of knowledge systems is necessary because farmers are much more accurate with their nomenclature in identifying soils and their suitability for varied uses.

Indigenous technical knowledge is the systematic body of knowledge acquired by the local people through accumulation of experiences, informal experiments and intimate understanding of the environment.

Over the years the impact of traditional and indigenous knowledge on agriculture has diminished due to the introduction of the synthetic chemical fertilisation system during the green revolution to attain the so called self-sufficiency in food production. But the consequences of dousing the soil with these deadly chemicals has been observed in the form of deteriorating soil fertility, contamination of the natural resource base and an increase in crop pest and disease incidence. As a result, an array of problems is being faced on the environmental, ecological and the health fronts by food producers as well as consumers. Eventually a revival of the sustainable, ecologically safe and socially sound practices is being sought in the present times by recognising and retrieving the traditional wisdom in agriculture. Complete and holistic technical knowledge is available about all aspects of cultivation of plants in our ancient literature The following traditional indigenous farming practices need to be now noted:

### Soil Fertility/Nutrient Management

Manuring or maintaining the soil's fertility has been termed the most important activity in cultivation. The Krishi-Parashara states that crops grown without manure will not give yield and the method of preparing manure from cow dung is described. Kautilya mentions the use of cow dung, animal bones, fish and milk as manure. The Kural (first century AD) observes that manuring is more beneficial than plowing.

During the pre-green revolution era, soil health and productivity was sustained through practices like crop rotations with legumes, intercropping/mixed cropping and the use of green manures. The practice of alternating the main crop with a leguminous one helped restore the productivity of the cropping system in the long run.

### Composed Manure

#### *Open pit compost*

Preparation of farm yard manure (FYM) through open pits has been a common practice of preparing manure for the soil in India from time immemorial. Animal dung, kitchen wastes and other organic farm wastes are collected in a pit and allowed to decompose to become farm yard manure over a period of time. In many places farmers adopt a two pit system which helps in removal of the manure alternatively. The sweepings from the cattle shed containing finely ground wheat straw or paddy husk is also added to the compost pit.

Kitchen wastes, ashes from the kitchen, grass clippings, green succulent weeds and other leafy materials are also collected and put into the pit. The dimensions of the pit were very much different from the types being made in present day times: the depth of the pit used to be around 2–2.5 metre and the pit used to be topped to the extent that the waste materials stood 30 cm above its edges.

*Animal hay-bed compost*

Another form of compost is animal hay-bed manure. Its practice is usually to be found in the northern states. In this method, the animals are kept an open space of the farm and fed with fodder that will contain hays of leguminous and other crops. The animals eat away the softer parts, leaving the hard stems behind. These form a hay-bed on the floor which is then partly crushed by the animals' movements and partly also cut into smaller pieces by their hoofs. At the same time, as the animals excrete and defecate, their dung and urine also get thoroughly mixed with the hay-bed. The resultant mixture is ready for use in five to six months. It is then removed and applied to the fields before the commencement of the rainy season. After it has been spread evenly on the surface, it is worked into the soil with the plow.

The water hyacinth (*Eichhornea crussipes*) is an aquatic weed commonly found in lakes, ponds, tanks and small ditches. It has been found to act as an effective bioremediant for heavy metals. Farmers collect this weed and heap it in nearby pits where it is allowed to decompose for nearly 30–40 days. After that the compost is sun dried, applied to plots and then incorporated into the soil through ploughing.

Biogas slurry is a by-product of biogas plants. This input is available in almost all Indian villages as biogas is used for cooking. Biogas slurry is used as an important input for the growing of vegetables like brinjal, chili, tomato, ladies finger, etc. It is estimated that about 28 million tonnes of biogas slurry are available in the country.

Tank silt application is the common practice followed by rice growers in the western states of India. There are a number of small, medium and large sised tanks and ponds in these states. During the summer months, when most of these tanks/ponds go dry, the farmers desilt them to harvest the dried or the semidried silts. These are then applied directly to the rice fields by summer ploughing. The beneficial effect of tank silt has been well documented.

The practice of penning cattle in fields left fallow is a common practice in the villages of India. Wooden logs are mostly used to tie down the animals in the fields. These logs are periodically shifted so that the whole field eventually gets a uniform distribution of animal dung and urine. Cattle herders who used to own large herds of sheep, goat and cattle were requested by farmers to leave their animals for the night in their fields for which they were paid. This practice enhances the process of on-site decomposition of animal feed residues and provides the nutrients to the soil directly without any losses. The practice of cattle penning is adopted in almost all parts of the country.

Human excreta is a potential source of nutrients. It was a common practice in earlier days to add night soil directly to the fields directly in areas adjoining the immediate periphery of the villages. The land so treated used to be considered the most fertile in the village (in eastern Uttar Pradesh it was generally called 'goin land'. The highest yields were obtained from these fields. The use of night soil, however, is not permitted as per current national and international organic standards.

### *Green manuring*

A variety of crops were used by farmers in different regions of the country for the purpose of green manuring. Following this practice, a farmer sows a crop and allows it to grow to certain stage before it is ploughed back into the field. In the northern parts of the country, crops like cowpea, sesbania, mung-bean and cluster beans are generally used for green manuring. With these crops, the farmers first harvest the beans and then plough the rest of the plants into the field. This routine provides them with an additional source of income as well as good manure for the soil. In the southern parts of country, crops like cowpea, black gram and gliricidia sp. are used as green manure. This increases the nutrient content of the soil and also helps with the control of weeds. In the western part of the country, green gram is used as a green manuring crop.

Application of thick green leaves is also a practice under green manuring in southern India. Leaves from the mango, jack fruit, cashew trees, etc., are preferred as they have long lasting residual effects. In western India, the leaves of jute are allowed to dry when they get detached from the stem. The dried leaves are incorporated into the soil to decompose before the sowing of the next crop.

### Fallowing

Leaving the land fallow between the taking of two crops was another traditional practice. Farmers manured the field and allowed the compost to mix with the soil and decompose thoroughly in order to encourage the mineralisation of the nitrogen. This practice was observed to enhance invariably the yields of the consequent crop.

### Pest Management

Reliance on cow dung for insect and disease control has also been reported in addition to its use in manure preparation. The use of cow dung for treating seeds, dressing wounds, plastering cut ends of vegetatively propagated crops such as sugarcane setts, sprinkled as a diluted suspension on plants and applied to soil has been indicated since the time of Kautilya. For instance, Kautilya's Arthasastra mentions that the use of a mixture of honey, lard and ghee followed the application of cow dung to seal the cut ends of sugarcane setts. Honey is mildly antimicrobial; ghee could seal off cut ends from contaminated soil as well as prevent loss of moisture from the setts, and cow dung could facilitate the biocontrol of potential pathogens.

A variety of traditional biopesticides has been used in the past to combat pest-problems in plants. These biopesticides are natural plant products that belong to the so-called secondary metabolites and which include alkaloids, phenolics, etc. These substances usually do not have any known function in photosynthesis; however, their action against certain other organisms is established. The plant kingdom offers us an array of such chemicals. About 2,121 plant species have been reported to possess pest control properties; 25 of these plant species possess characteristics required for an ideal botanical insecticide. We can now discuss plants that are most widely used for pest management and control.

### Insect Pest Management

#### *Adath odavesica*

One kilo of fresh leaves is finely pounded, mixed with six litres of water and allowed to soak for a full day. The liquid is then filtered and a suitable sticker, such as soap, is added. Applications are best done in the evening time after sunset. The mode of action is antifeedant, insecticidal and acaricidal.

#### *Azadirachta indica (neem)*

Neem has emerged as the single most important source of botanical insecticides. Neem leaves and seeds contain azadirachtin, miliantriol, etc., having insecticidal properties. About a hundred protolimonoides, limonoides or tetranor-triterpenoides and some non-terpinoides have been isolated from various parts of the neem tree. Neem products are also used as nitrate inhibitor. More than 300 species of insects have been reported to be affected by neem compounds. Farmers in India were aware of the insecticidal properties of neem even before the invention of chemical pesticides.

Some of the important preparations of neem are as follows:

*Neem Seed Kernels Extract (NSKE)*

NSKE is prepared from dried neem seeds. About 25 kg of dried neem seeds are collected and crushed into coarse pieces. These crushed seeds are tied in a muslin cloth and soaked overnight in a container having 50 litres of water. The cloth containing the seeds is squeezed to discharge the extract. The tied seeds are again dipped in the water and squeezed a second time to extract the whitish neem seed kernel extract completely. Now good quality water is added to the concentrated NSKE solution. A spray solution of 5% can be used to repel insect pests, if any. For uniform spreading an emulsifier is usually added. A locally made soap can be used for this purpose. This extract is also used as a prophylactic measure. It has to be used up within 1–2 days of being made to avoid degradation of the active ingredient.

*Neem decoction*

For making a neem decoction, about 20 gm of kernels are taken and tied in a cotton cloth (30 gm of neem seed is required to obtain 20 gm of kernels) and kept in water for about 6–10 hours. After this the bundle is taken out and squeezed to remove the extract. The extract first coming out will resemble rice water mixed with turmeric powder. The squeezing is continued till the colour of the extract is clear. This extract can be used as a spray for the plants.

*Neem oil*

Neem oil can be used as a good insecticide. The spray solution is made by mixing the neem oil in a soap solution.

For making one litre of solution, 20 ml of neem oil is mixed in one litre of water. Thin slices of soap are added to this solution and mixed properly till white foam is produced. (Instead of soap slices teepol can also be used.) The incidence of stem borer in sugarcane can be reduced by the use of neem oil.

*Neem cake*

Neem cake can be used for controlling nematodes. Approximately 50 g of neem cake should be applied in the pits before seedlings are planted. A spray solution can also be prepared from the cake; for this, one kilo neem cake is soaked overnight in five litres of water. The solution is filtered and then sprayed.

*Neem leaf decoction*

Neem leaf decoction can be made by boiling neem leaves in water. About 6–8 kg of fresh neem leaves are plucked, put in 10 litres of water and boiled till the solution attains a dark brown colour. After 10 to 12 hours, the solution is mixed with 80 to 100 litres of clean water and then sprayed as per requirement. The dried leaves of neem are also put to use for the control of storage pests in wheat. In certain areas, a paste made with neem leaves and water (10 kg : 2 litres) is used for effective control of the leaf folder in rice.

*Acorus calamus*

Rhizomes of this plant contain calamus, which has insecticidal properties. Saponins and tannins are the main active principles. Pest control formulations are made using dried and powdered rhizomes. In addition, aqueous extract can also be made in solvents like petroleum ether, kerosene or ethyl ether for better spread and efficiency. The dried slices of the rhizome can be kept with stored products to ward off storage pests.

### *Annona squamosa*

The leaves and seeds of this plant contain lanolin and anonaine having insecticidal properties. The seeds of the plant are used for making botanicals. The plant is useful against stem borers, fruit borers, sucking pests, scale insects and hoppers.

### *Turmeric (Curcuma longa)*

Turmeric powder is being used by farmers to control tomato wilt. A solution of turmeric powder is used to treat the roots of seedlings in tomato nurseries before they are transplanted. Another turmeric solution adopted by some farmers is for effective control of the cotton semi looper, mites, green leaf hoppers, aphids, etc. This solution is also effective to control some diseases of plants. It is prepared by powdering 1 kg of dried turmeric into a fine powder which is then added to 3 to 4 litres of cow urine. The solution is properly mixed and 4 ml/litre of soap solution is added to improve its stickiness.

### *Asafoetida*

Asafoetida is used by farmers for the control of termites (*Odontotermes obesus*). About 40–50 gm of asafoetida is packed into a cotton cloth. Two to three such packs are placed in the irrigation channel at a distance of about 20–30 m from each other during irrigation. Asafoetida solution is also used by some farmers for the control of larvae of Heliothis. About 100 to 150 gm of asafoetida (per litre of water) are boiled for 10 to 15 minutes. After the solution is cooled, it is sprayed with 40 to 50 litres of fresh water per hectare.

### *Onion* (Allium cepa)

Farmers have reported that the use of onion along with common salt is effective for the control of storage pests.

About 2–3 kg of common salt and 3–4 kg of onion are mixed for every 100 kg of chickpea (*Cicer arietinum*) seed in storage.

### *Capsicum*

Use of dry chili (*Capsicum annuum*) is one of the most popular practices for controlling insects that invade storage. For instance, for the control of storage pests in mung bean (*Vigna radiata*), farmers place 4–5 dry chilies per kilo of seed. Known to be quite effective.

### *Calotropis gigantea*

Calotropis can be used for control of termites and for caterpillars of insects. For termite control, about 8–10 kg of the plant material is soaked in water for 24 hours. The liquid is filtered and poured over the termite affected soil. Farmers also use an indigenous method of testing the effectiveness of the solution: they place pieces of wood in the soil at various points; if the wood is free from termites, the solution is judged to be effective.

Another solution can be made also from the latex of the same plant. The latex of calotropis is diluted with water at the rate of 1 :15. This effectively controls caterpillars of insects within three days. A liquid preparation can be made by putting 1 kg of calotropis leaves in 5 litres of water and leaving the mixture for a month. The solution is then filtered and sprayed: it is effective in controlling leaf sucking pests.

### *Castor bean (Ricinus communis)*

About 500 gm of shelled castor seed or 750 gm of unshelled seed are first crushed. The crushed seed is heated in two litres of water for ten minutes. A little soap and two table spoons of kerosene are added. The solution is filtered and then ten litres of water is added into it. The final solution is sprayed on to the plants. Use of kerosene is not permitted under national and international organic standards.

### *Nicotine*

Nicotine is a quick acting insecticide. It is also effective in its gaseous form. Nicotine is an alkaloid; in liquid form, it does not have any colour or smell but when it comes in contact with air, it turns black. It disappears after application. Therefore, nicotine sulphate, which is available in 40% liquid form, is used for insect control. It can be diluted with water and mixed with soap and sprayed on to the plants. Tobacco use is not allowed under existing organic standards.

### *Bidens pilosa (Spanish needles)*

The whole plant and its seeds can be used to make an insecticide which can help in controlling aphids, cutworms and termites. For preparing a formulation, mature seeds are first boiled in water for ten minutes. After cooling, a little more water is added along with a piece of soap. The solution is then filtered and sprayed. The whole plant can also be used by pounding it and soaking it in water overnight. A small piece of soap is added to it before it is sprayed.

### *Lantana camara*

A very effective liquid manure can be prepared by utilising the whole plant. In addition, water extract of its dried leaves and flowers is used to control many insects including aphids.

### *Coriander*

About 200 gm of crushed seed are boiled for ten min in one litre of water, filtered, and diluted at a rate of 1 : 2 with hot water. This acts as a good repellent against spider mites.

*Eupat orium sp (Siam weed)*

A very effective repellant can be made from this plant for help in the control of aphids, diamond back moth, snails and storage pests. Four hundred grams of dried leaves are boiled with ten litres of water for ten minutes. After cooling, the solution is filtered and sprayed.

*Rotenone*

The roots of the plant Derris elliptica, a leguminous plant, contain 4–11 per cent rotenone. For their utilisation, they can either be dried and powdered or mixed with water. Rotenone is useful against sucking pests, caterpillars and some beetles. It acts as a contact and stomach poison. Use of rotenone is not permitted under national and international organic standards.

*Ocimum basilicum*

This plant is rich in alkaloids. All parts of the plant including its stem, leaves and roots can be dried and the alkaloid content extracted by using solvents like ethanol or acetone. Even water can be used very well for the purpose. The solution can help in the control of flies, mosquitoes and nematodes. The leaves of the plant can be boiled with water and sprayed after adding soap solution. This is done at the rate of 100 gm per one litre of water with 1 ml of soft soap. The solution can be used for effective control of aphids, army worm, red cotton bug, cockroach, mosquito, etc.

*Artemisia*

An extract of this plant is effective as a spray against common insects. The whole plant is harvested and the leaves and stems are cut into small pieces. One kilo of the plant material is soaked in ten litres of water for 24 hrs until the fermentation commences. The fermented decoction is filtered to obtain a clear liquid before it is sprayed.

### *Tobacco tea*

Tobacco tea is prepared from tobacco leaf and is often used against thrips, scale insects, mealy bugs, aphids, beetles, caterpillars, etc. About 400 gm of tobacco leaves are cut into small pieces and soaked in four litres of water for one day. They are then heated a little and pressed. Afterwards, the solution is filtered to obtain a clear liquid. Soft soap is added @ 1 ml per litre to this solution and it is stirred thoroughly. The concentrated decoction is diluted 5 to 6 times before spraying. Use of tobacco is not permitted under national and international organic standards.

### *Ipomea fistula*

Farmers collect 1000–1200 fresh leaves with buds from the hedge of Ipomea spp. The leaves are put in 30–35 litres of water and boiled until the water turns milky white. The solution is used to control the larvae of Heliocoverpa armigera, spotted bollworm and army worm. Farmers have reported that this solution has to be used within four days from its preparation.

### *Pyrethrum*

Pyrethrum is extracted from the dried flowers of Crysanthemum cinerariafolium. Pyrethrum or natural pyrethroid is a mixture of six different substances. The main active ingredient is the alkaloid stachydrine. Pyrethrum can be effectively used to control bugs, ants, flies, cockroaches, thrips, scale insects, mealy bugs, etc. It is also useful against the storage insects.

### *Aegle marmelos*

The fruits of this plant are used to make an extract that is effective against thrips. About 20 to 25 fruits are boiled in ten litres of water to make the extract. One kilo of turmeric is added and the solvent kept for twelve hours. One litre of the solution is diluted with ten litres of water and is sprayed over the entire plant.

*Marigolds*

The extract of marigold flowers has been known for its use against insect pests and nematodes. The flowers are crushed to a paste and diluted with water to make sprays. Growing of marigolds is recommended in kitchen gardens and around homes to repel mosquitoes.

*Other formulations*

Other plants that can be used: Ryania speciosa contains the alkaloid ryanodine and is effective against lepidopteran pests; datura is an effective antifeedant; mushrooms contain strobllurin which inhibits spore germination. A few microorganisms like Phytopthora palmivora, Colletotrichum gleosporides and Streptomyces hygroscopius are also used as bio-weedicides.

— A solution prepared from 150 ml of lemon juice and 150 ml of tamarind juice in 15 litres of water is effective for control of hairy caterpillars. The solution prepared with those quantities is sufficient for a 0.25 hectare area.

— Equal quantities of chilies and garlic are crushed together and water extract taken. Diluted with 200 litres of water, the extract is effective against aphids and jassids.

— Tobacco wastes are used to make an extract which, with 250 gm of soap solution in 200 litres of water, is used as a spray to control stem borers.

— A mixture of 10 kg ash and 1 kg sugar is used to control termites. The mixture is kept at the place where termites are noticed. It will attract ants which will feed on the termites.

— Seeds mixed with Acorus calamus (baje) powder in the ratio of 10 kg : 1 kg respectively will remain free of storage insects for long periods.
— Neem leaf powder in a concentration of 1% is used to keep green gram free from any insect infestations.
— Turmeric powder is used as a repellant to control insects in stored grain.
— Farmers smear seed of wheat and other crops with castor oil for long term preservation.
— A layer of 5–7 cm sand is effective in protecting pulses from beetle attack.
— Pulses mixed with the cow dung powder and kept in earthen pots can be stored for considerable periods.
— Larvae of Heliothis are boiled to make a dark brown coloured solution which is used as a repellant against the same larvae. When sprayed, the solution renders the field free from the larvae.

All the practices discussed above are adopted by farmers. The source of these practices is the knowledge inherited from their fathers and forefathers. These practices can be considered to be time-tested and effective.

**Plant Disease Management**

India is a veritable treasure trove of indigenous traditional knowledge (ITK) concerning plant health protection and disease management that has evolved and also been tested for several centuries. In general, there are about a thousand diseases that can potentially affect economically significant crop plants by causing major damage. Almost every crop is attacked by at least one or more diseases. About 800 diseases are caused by fungi alone while approximately 50 each are due to bacterial, viral and nematodal vectors. A few are due to phytoplasmas.

The further development and adoption of traditional knowledge is necessary considering that the diseases prevalent now may not have been present in earlier periods. However, the basis of the methods adopted has common characteristics such as:

1) Multi-portaged attack on the pest/disease.
2) Focus on improving plant health thereby increasing its resistance capacity.
3) Enriching soil with (and buffering) beneficial microbial activity.
4) Broad spectrum effects on pest and diseases which is desirable.

In the ancient literature of Surapala's *Vrikshayuraveda*, the basic concept of disease identification in plants is based on the *tridoshas* i.e. *vatta, pitta and kapha*.

A variety of methods have been suggested for the treatment of disease. These include treatment based on mantras; practices using organic materials; practices using inorganic materials; and use of other products of animal origin. The oldest document reporting the use of organic materials to control crop disorders is probably Kautilya's *Arthashastra*. Other products used in the management of disease are milk, ghee, honey and buttermilk.

The traditional practices had a two-point strategy for combating disease problems in crops. These were (i) disease avoidance and (ii) disease management.

The second strategy we have mentioned is the management of disease. Traditional practices adopted to manage disease include the following:

1) A common practice in the Western Ghats and Konkan region to combat the soil borne pathogens is nursery burning with leaf litter and farm wastes.
2) Another practice to manage some soil borne pathogens like pythium, phytophthora, etc., is the use of raised beds, fields and ridges.

3) Flooding with water is followed to control the problem of soil borne pathogens.
4) The problem of pythium damping off in nurseries of brinjal and tomato is overcome by earthing up.
5) Seed borne diseases in wheat like wheat bunt are treated with salt water.
6) Appropriate planting/sowing time helps to overcome the problem of some foliar diseases.
7) Use of tobacco: Tobacco has been used traditionally to control mosaic and bacterial wilt in tomato and brinjal. The preparation is made by mixing and grinding one kg of tobacco (*Nicotiana tobaccum*) leaves in 15 litres of water. The solution is left untouched for 2–3 days and then filtered and sprayed on the plants. Use of tobacco is not permitted under national and international organic standards.
8) Farmers' practices of summer ploughing and crop rotation with legumes, cereals and millets reduce the problem of soil borne pathogens.
9) Sterility mosaic of pigeon pea is managed by mixed cropping of jowar with tur which inhibits movement of vector mites.
10) Cow's urine: Application of cow urine prevents bacterial wilt (*Pseudomonas solanacearum*) in tomato, potato and chilies. Cow urine is left undisturbed for two weeks and allowed to ferment. This fermented urine is diluted with water and sprayed on the plants.
11) Application of organic manures also helps to manage soil borne pathogens. Good quality manure contains an appropriate balance of all the microorganisms necessary to provide the soil with defences against pathogens.
12) Airborne diseases are overcome by planting across the wind direction.

13) Asafoetida and turmeric: Asafoetida and turmeric have been traditionally used to prevent bacterial wilt in tomato and brinjal. The soil is drenched with a solution prepared by adding one gram of asafoetida and 5 gm of turmeric in ten litres of water.

In addition to the use of specific plants or inputs for the management of disease, there are certain indigenous practices or preparations manufactured by farmers from locally available materials of daily use, which are very effective in controlling disease in plants.

— Milk solution (one litre of milk in nine litres of water) is effective in the control of powdery mildew and viral diseases of brinjal, tobacco, potato, tomato, chilies, etc.
— Extract of leaves of Casuarina equisitifolia is effective to control some bacterial and fungal diseases. About 20 kg of its leaves are boiled in water for 20 minutes. After cooling, the solution is filtered and then diluted and sprayed. The extract obtained is sufficient for an acre.
— A solution of papaya leaves is effective in the control of brown leaf spot disease in paddy. Approximately 2 kg of fresh leaves are soaked in 3 to 4 litres of water overnight. The solution is filtered and diluted with 50 to 60 litres of water. About 250 ml of soap solution is added before spraying.
— Wheat seeds are soaked in milk before sowing to avoid rust disease.
— A solution prepared from 1 litre of buttermilk in 20 litres of water is effective to control leaf curl in tobacco.

**Crop Improvement**

In addition to the problems of insects in field and storage, there are certain other difficulties faced by farmers like seed dormancy, the problem of non-bearing, low yield, reduced flowering, etc. Certain practices have been developed by the farmers to combat these situations. These practices are discussed below:

*Flowering:* Farmers use a 5% solution of jaggery to improve flowering and fruiting in cucurbits such as bottle gourd, pumpkin and watermelon. Some farmers place the lower half of the fruit e.g., the bottle gourd, in pots containing 5% jaggery solution for a week to increase its size.

*Breaking non-bearing nature:*Farmers apply fish compost to the roots of orange and lemon to break their non-bearing tendency. Withholding of irrigation in deep vertisols (which generates stress) is still practised. Also spraying of 5% solution of cane sugar on the leaves is another interesting practice in vogue. The non-bearing nature of papaya is generally cured by fixing a piece of iron at two or three places in the stem. Farmers report that this is the only way to break the non-bearing nature of papaya. The practice restricts the flow of water and nutrients and thus induces the reproductive phase.

*Enhancing yield:* Farmers irrigate bottle gourd with the water of cooked rice for good yield. They apply about 1 to 2 litres in a day thrice a week. The growth in the fruits of bottle gourd is governed by osmotic pressure.

*Breaking dormacy:* Farmers put water melon seeds in a lump of cow dung and keep it near the fireplace used to cook food. After three days, the seed is removed and sown. Similarly, farmers are known to cut the primary roots of brinjal seedlings to control a potential infertility problem. In another practice, farmers break coriander seeds with moist leather shoes to avoid manual damage. This practice is done to split the seed without damaging the embryo. Breaking of coriander seed is necessary for good germination. Immersing seeds of bitter gourd, onion and pumpkin in moist ash of burnt wood or cow dung overnight is also practiced to enhance their germination.

Likewise, sugarcane setts are put into the waters of a pond for the same purpose. The action facilitates the conversion of sucrose into glucose and fructose as energy sources.

**REFERENCES**

Altieri. M., "The green revolution revisited: New needs, new strategies", Ceres (154), 1995.

Arunachalam V., "The science behind tradition", *Current Science*, 2001.

Fukuoka, M., "The natural way of farming: The theory and practice of green philosophy", Bookventure Publishers, Madras, 1997.

# 3

# Inputs for Organic Farming

Plants are autotrophic and have the unique ability to extract their nutrients from two sources viz., atmosphere and soil. There are thirty-four elements are necessary to plants and these are divided into two large groups viz., those derived from the atmosphere and those derived from the soil.

*Table 1. The atomic elements that constitute living organisms or participate in their metabolism.*

| *Elements from the atmosphere* | *Elements from the soil* |
|---|---|
| These represent 92 to 98% of a plant's dry weight | They represent 2–8% of a plant's dry weight 12 Vital elements of which; |
| 4 Constitutive elements | 2 are non-constitutive – K and Cl |
| C, O, H and N | 10 are constitutive – P, B, Ca, Mg, S, Fe, Mn, Mo, Cu and Zn |
| | 18 other elements (functions of some are not fully understood) |

This table indicates that the plant feeds quantitatively for the atmosphere and qualitatively from the soil. A very few atmospheric elements make up the bulk of dry weight or vegetal matter and any soil elements constitute only a very little part of the vegetal matter. The soil provides two types of elements: non-constitutive elements which the plants borrows and returns to the soil at the end of the period of cultivation; and constitutive elements which enter the vegetative organic molecular composition.

Carbon and oxygen together constituting 88% of plant tissue are derived from carbon dioxide. Photosynthesis allows the integration of carbon and oxygen from carbon dioxide within the structure of a sugar. Hydrogen, which contributes 6% of plant tissues, is derived from the breakdown of rainwater and hydrogen, which is absorbed in the tissue of the plant. Nitrogen is derived from the atmosphere, through microbes that can transform gaseous nitrogen into ammonia. The ammonia directly enters in leguminous plants in the form of amino acids or is converted into nitrates by the soil microbes and then absorbed by other plants.

The constitutive elements of the plants derived from the soil do not enter in the form of atoms but in the form of oxides and chelates. These transformations are affected by the soil microbes, which render these elements accessible to plants from the parent rock or humus. Thus all other soil elements can be absorbed by plants through microbial action. The oxidation of soil elements is often achieved by very specific micro flora. Fifteen oligo-elements can enter plants in the organic form. Similar studies were reported by Milton Whitney, Chief Bureau of Soils, USDA who concluded that 95% of the constituents of the dry vegetation are from the atmosphere and 5% is from the soil. The statistics clearly reveal that the plants do not have a purely mineral diet.

'Agriculture research,' says Dr. Albert Howard, 'has been misused to make the farmer, not a better producer of food, but a more expert bandit. He has been taught to profiteer at the expense of posterity –how to transfer capital in the shape of fertility and the reserves of his livestock to his profit and loss account. In business such practices end in bankruptcy; in agricultural research they lead to temporary success. All goes well as long as the soil can be made to yield a crop. But soil fertility does not last forever; eventually the land is worn out; real farming dies.'

Considerable quantities of organic resources are available for recycling from agriculture and its allied activities. Organic matter is an important soil constituent influencing favourably a number of constraints linked with crop productivity. It is widely accepted that high levels of soil organic matter (SOM) mean high potential productivity and soil health.

In India, soil organic matter varies form 0.05% to 5%. Even at low concentrations, SOM is the major substance facilitating aggregation and stability of soils. Apart from contributing adequate quantities of plant nutrients, SOM offers a platform for a variety of beneficial microflora and fauna to thrive upon. If these resources are used appropriately and scientifically, they can open new vistas in the utilisation of the yet untapped source of nutrients. The potential sources of plant nutrients are:

a) Soil reserves

b) Livestock wastes
   - Cattle wastes
   - Poultry waste
   - Swine waste
   - Goat and sheep excreta

c) Agricultural residues
   - Aquatic weeds
   - Forest litter

d) Agro industrial wastes
   - Municipal solid wastes (MSW)
   - Oil cakes
   - Sugar industry wastes
   - Vegetable and fruit processing wastes

e) Fisheries and marine wastes

f) Marine algae and sea weed

## SOIL RESERVES

As a reservoir of nutrients, the soil provides all the nutrients for the growth and development of plants, akin to a forest ecosystem. Studies suggest that fertile soils in the top 0–30 cm contain 20,000 kgs.ha-1 of nitrogen, 30,000 kgs.ha-1 potassium and 50,000 kgs.ha-1 of calcium. Continuous cropping over several years without recycling has resulted in the depletion of plant nutrients and of the symbiotic association of the micro flora and fauna. It has been estimated that soil depletion of N, P and K is 8–10 million tonnes (mt) per annum at the national level. The deficiencies of N, P, K, S, Zn, Fe and Mn are common in many soils in India.

### Livestock Waste

#### *Cattle wastes*

The total cattle population in India is estimated at 283 million. An average dairy cow produces between 10 and 18 tonnes of dung and urine per year. In each tonne of dung, about 8–12 kg of nitrogen (N), 2–4 kg of phosphorus ($P_2O_5$), 3–6 kg of ($K_2O$) is available on an average. However in India the value of dung as manure is reduced because of run off, leaching and volatilisation of nitrogen. It is estimated that during the winter months, 40% of the collected dung is dried into dung cakes and 58% put into compost pits and heaps. However, during the rainy season only about 13% is dried into dung cakes and 85% find its way into manure. About 2% of cattle dung is used for other household purposes such as plastering of floors and walls. The dung collected for fuel purposes was estimated at 30% in the early 1970s and up to 75% in the early 1990s. Cattle waste provides tremendous potential for its utilisation in agriculture as farm yard manure and compost and in the form of cow urine as a spray for nutrition and pest management.

### *Poultry waste*

Large quantities of wastes in the form of litter (bedding materials) and droppings are generated from poultry units. Poultry waste is rich organic manure. The excreta of a hen 1–6 months old ranges from 250–800 grams per head per day. The poultry population India is 490 million. Poultry litter comprises 2.5–3.0% nitrogen, 0.6–1.13% phosphorus and 1.0 to 1.8% potassium.

### *Swine waste*

Waste production from piggery units comprises mainly of pig dung, unfed material and urine. The quantity of wastes depends upon the type and size of animal. Pig manure has about 0.53% nitrogen, 0.53% phosphorus and 0.36% potassium. Approximately 30% of the consumed feed is converted to body tissue and the remainder is excreted as urine and manure. In India, the swine population is about 14 million.

### *Goat and sheep excreta*

The goat with a body weight of 20–40 kg excretes 0.3 to 0.6 kg dung and 0.3 to 0.5 litres of urine, whereas the sheep with a body weight of 25–40 kg excretes 0.3 to 1.4 kg dung and 0.3 to 1 litres of urine per head per day. Sheep and goat manure contains about 3% of nitrogen, 0.4% phosphorus and 1.7% potassium.

## AGRICULTURAL RESIDUES

The term 'agricultural residues' is used in a broad sense to include wastes from agriculture and agro-industries. They may consist of unutilised excess materials, or of residues from the growing and processing of raw agricultural products such as fruits, vegetables, grain, and trees. Residues are those end-products of production and consumption that have not been used, recycled or salvaged.

They are the non-product flows of materials and energy whose economic values, at the present level of knowledge, are less than the cost of collection and transformation for use, and they are therefore discarded as wastes. The volume and composition of these residues could be reduced if there were technological means for converting them into some usable product and if the value of the subsequent product were to exceed the costs of conversion.

If residues can be utilised for human benefit, they are no longer wastes but become new resources to augment limited existing ones. Residue utilisation is resource utilisation. Recycling, reprocessing, and utilisation of all or a portion of the wastes offer the possibility of bringing such residues into beneficial use. The key to successful residue utilisation is the suitable technology to process the residue under local conditions. The need is to consider residues as potential resources rather than as undesirable wastes. Crop residues represent about half the nutrients exported through the production of grain, fibre or nut crops.

Thus, the better use of residues from agricultural production can help to accomplish two important objectives: (a) better utilisation of existing resources to meet human and animal food needs, and (b) reduction of potential environmental problems caused by such residues.

*Table 2: Approximate quantity of crop residues and their NPK value*

| *Crop residues* | *Quantity tonnes/ha* | *N* | *P (%)* | *K* |
|---|---|---|---|---|
| Mustard straw | 1.8–2.2 | 0.45 | 0.53 | 1.48 |
| Castor shell | 0.5 | 1.01 | 0.35 | 1.86 |
| Wheat straw | 1.8–2.5 | 0.53 | 0.10 | 1.10 |
| Cotton stalk | 4.0–5.0 | 0.44 | 0.10 | 0.66 |
| Paddy straw | 2.20 | 0.36 | 0.08 | 0.71 |
| Rice husk | 0.25 | 0.45 | 0.25 | 0.45 |
| Sugarcane trash | 9.0 | 0.35 | 0.12 | 0.60 |

| | | | | |
|---|---|---|---|---|
| Maize cobs | 0.8–1.2 | 0.42 | 1.57 | 1.65 |
| Groundnut shell | 0.8 | 1.25 | 0.46 | 1.50 |
| Banana pseudostem | 9.0 | 0.61 | 0.12 | 1.0 |
| Pigeon pea stalk | 1.5–2.0 | 1.10 | 0.58 | 1.28 |
| Pearl millet straw | 0.4 | 0.65 | 0.75 | 2.50 |

Water hyacinth (Eichhornia crassipes) is a free floating weed plant that grows luxuriantly in ponds, lakes and water reservoirs. Its rapid growth has adverse effects in agriculture, fisheries, water transport and human health. However, there is great scope to use it as soil mulch, green manure and compost.

Forests provide abundant leaf litter. It is estimated that more than 15 million tonnes of compost can be obtained from forest litter annually without adversely affecting the natural regeneration of the forests. In the Dhamtari forest division of Chattisgarh state, attempts have been made to use the leaf litter for composting with the help of the tribals.

## SOIL FERTILITY IMPROVEMENT

A number of other organic inputs can also be used to boost the growth of crops, improve yields and the quality of produce. Farming using organic inputs is profitable because all such inputs can be prepared on the farm itself using locally available resources and the costs of production are low.

### Biofertilisers

Biofertilisers are living microorganisms of bacterial, fungal and algal origin working singly or in synergy with one another in a consortium which can be used to increase crop productivity. These microorganisms assist in biological nitrogen fixation, solubilisation of insoluble phosphates in the soil, stimulation of plant growth and control of diseases.

Biofertilisers are apparently environmental and farmer friendly renewable sources. They are non-bulky, low cost agricultural inputs playing a vital role in improving the nutrients availability to crop plants. Today, a diverse variety of bio-inoculants such as nitrogen fixers, phosphate solubilisers and plant growth promoters are used in agriculture.

### *Rhizobium*

A rhizobium is a symbiotic bacterium which colonises the roots of specific legumes to form root nodules with the capacity to fix atmospheric nitrogen. Rhizobia have been widely and extensively used as a biofertiliser because of their proven ability to fix and contribute nitrogen for the benefit of leguminous host plants. The rhizobium legume association can add up to 20–25 kg nitrogen per hectare and leave a substantial amount of nitrogen for the succeeding crop.

### *Azotobacter and Azospirillum*

These are non-symbiotic bacteria and are capable of contributing nitrogen to non-leguminous crops, mainly cereals and millets.

Azotobacter is a free-living heterotrophic nitrogen fixing bacterium found in neutral alkaline soils. These cultures can be used for cereal crops like rice, wheat, barley, sorghum, maize and vegetable crops like tomato and potato. They are also used in crops like mustard and cotton.

Treatment of seeds and seedling with azotobacter culture is a very simple process. Besides their ability to fix nitrogen from air, they also secrete growth promoting substances which help in better seed germination and expanded root systems which facilitate the plant's nutrient uptake.

Azospirillum is an aerophillic nitrogen fixer, commonly found in association with the roots of cereals and grasses. High nitrogen fixation capacity, low energy requirement, abundant establishment in roots of cereals and tolerance to high soil temperature make it more suitable for use under tropical conditions. It can be of use for crops like sorghum, pearl millet, minor millets, barley and forage legumes. These bacteria can fix 20–40 kg N/ha.

### *Azolla*

Azolla is an aquatic fern widely found in natural conditions in temperate and tropical regions as well as in lowland rice-growing areas. The fern fixes atmospheric nitrogen in symbiotic association with Anabena algae. This association uses energy from photosynthesis to fix atmospheric nitrogen upto 100–150 kg per hectare annually. Azolla is widely used in paddy cultivation. It can be used either as a green manure before transplanting or as a dual crop after the transplanting of rice. The growth and efficiency of Azolla is generally better in hilly, humid and coastal regions than in high temperature regions.

### *Blue Green Algae (BGA)*

Blue green algae or cyanobacteria, the photosynthetic nitrogen fixing organisms, are important biofertilisers for rice cultivation. Studies reveal that BGA applied at the rate of 10 kg per hectare after seven days of transplanting of rice contributes about 20–30 kg nitrogen per hectare per season. Besides fixing nitrogen, these algae excrete vitamin B12, auxins and ascorbic acid all of which contribute to the growth of the plant. Application of algal biofertilisers is also useful in the reclamation of marginal soils such as saline-alkaline and calcareous soils.

### *Phosphate Solubilising Microorganisms*

Phosphorus is an important plant nutrient that is not

readily available. Some heterotrophic bacteria have the ability to solubilise inorganic phosphorus from insoluble sources by secreting organic acids. The important phosphate solubilising bacteria are Pseudomonas spp. and Bacillus spp. Several fungi species belonging to the genera Penicillium and Aspergillus are also involved in solubilisation of phosphates in soil. These microorganisms can be used for all crops and can solubilise 20–30% of insoluble phosphate in the soil.

### *Vesicular Arbuscular Mycorrhiza (VAM)*

Mycorrhiza is the symbiotic association of certain fungi with the roots of higher plants. VAM, a fungal biofertiliser, is the most promising among the mycorrhiza. It colonises the roots and increases the growth and yield of crops over a broad ecological range – from aquatic to desert environments.

VAM increases plant phosphorus uptake and the ability of plants to take in other nutrient elements like zinc, copper, potassium, sulphur, etc. It also improves the water retention capacity of plants and enhances their resistance against drought and disease. Mycorrhiza is a broad spectrum biofertiliser which is also suited for nursery, transplanted and plantation crops. VAM plays a very important role in forest ecology for sustainable production.

## COMPOSING

In forest ecosystems, the cycle of birth and death is a continuous process.The fallen leaves, branches and trees get degraded by a variety of microorganisms and insects present in the forest which convert these dead material into humus. Composting is a technology or means to bring natural forces back to the earth and thereby enliven the soil. It brings about a balance in nature by facilitating recycling.

Composting has been misunderstood and misinterpreted as the mere degradation of organic waste. In reality, composting is an active process to bring about the desired degradation in order to support an array of beneficial microorganisms.

It is not required that the microbes are inoculated into the compost, but rather, measures instead be taken to nurture the localised microbes of the region. In every gram of top soil, millions of bacteria, yeast, moulds, diatoms and other microbes thrive. Composting is therefore an enlivening process. The end product of compost i.e., humus, plays a vital role in the nutritional, physical and biological functions of the soil.

Humus improves the physical condition of the soil. It acts as a binding agent, and holding the soil particles together in irregular shaped grains or crumbs. A soil with well developed crumb structure manifests all the good qualities associated with good soil: it has good tilth; it enables percolation of irrigation and rain water; it keeps the soil loose, reducing its scope for compaction. It also holds nine times more water than sand and makes the water available to the plant when it needs it.

The transfer of nutrients from soil to plant is dependent on the pH of the soil. Humus acts as a buffer, maintaining optimal pH conditions in the soil. It thus aids the proper supply of nutrients.

Good compost comprises antibiotics and phenols that are absorbed by the plant and which in turn impart resistance to it against insects and disease. When optimum conditions are created in the soil, microbes start secreting growth hormones that are generally referred to as phytomins–akin to vitamins in animals. When there is a deficiency of phytomins, plants look weak, become susceptible to insects and disease. Phytomins help in developing a symbiotic relationship with microorganisms in the soil. Different methods of composting are now discussed below:

## Indigenous Methods

Here compost is made by throwing all crop residues including straw, sugarcane refuse, green leaves, chaff, rice hulls, coir pith, forest litter and other available wastes into a heap 1.0 to 1.5 metres height. A depression is made at the centre to collect rain water and the heap is left to degrade on its own for six months before it is utilised in the field.

### *Indore Method (aerobic process)*

In this method, a pit of three feet deep, 6–8 feet wide and with any convenient length is dug near a cattle shed in a site free from waterlogging. Farm wastes are shredded and used as a bedding material for cattle. After use, the wet beddings along with the dung are heaped in a layer (3–5 metres) in the pit. A sufficient quantity of cow dung is spread over this layer. Water is sprinkled to maintain optimum moisture. The process of layering and spreading cow dung is repeated so that the layers are formed one above the other till the residues reach one foot above the ground level. The pile is built up before the start of the rainy season or a protective shed may be erected over the pit. The compost heap is turned every fortnight and the compost by this method will be ready in 16 weeks.

### *Bangalore Method (anaerobic process)*

In this method greater conservation of nutrients is attempted. Pits of size 4.5x1.5x1.0 metres are prepared and the residues placed in them as outlined in the Indore process except that each heap in the pit is sealed by plastering with a layer of soil about 25 cm thick. This process aids in increasing the temperature inside the plastered pit due to anaerobic fermentation.

The compost will be ready in 32 weeks.

## Biodynamic Composting

Biodynamic agriculture emphasizes the production of good quality compost for producing quality food. Where active agriculture is being undertaken, application of compost is very important. Under the biodynamic composting system, a set of six preparations numbered from 502–507, prepared from different herbs, is used in homeopathic doses. These have unique properties of bringing in the desired degradation of the waste materials. The compost does not have any odour. These preparations stimulate different nutrient cycles and make these nutrients available in an assimilable way.

In India, due to different geo-climatic zones, the biomass available for composting generally differs. Appropriate composting methodologies are devised depending on the type of biomass and the crop under cultivation. For instance, for crops like vegetables, the quantum of nitrogen material (green biomass, animal manure) in the compost is increased to 75% and carbonaceous material (biomass) to 25%. Such compost – generally called bacterial compost – aids the growth and developments of vegetables.

On the other hand, for perennial crops like tea, coffee, cocoa and fruit trees, the ratio of nitrogenous material to carbonaceous material is kept at 60%. Such compost – generally fungal compost – gives good results with these perennial crops. Crops are like home pets. Some crops are very specific to a particular type of compost. For instance, good quality and tasty tomatoes can be harvested if the plants are fed with a compost made from old tomato plants. Tomatoes have a narcissistic predilection for growing on their own rotted remains. Likewise, hog manure is best for root crops, cow manures for most crops, etc.

Composting can be carried out by following the traditional static pile method in which waste materials are layered one above the other and cattle dung slurry is sprinkled in between. The heap (optimum height of one metre) thus created is plastered with soil and cattle dung.

*Table 3: Role of biodynamic (BD) preparations*

| *Preparation* | *Activate the availability* |
|---|---|
| BD502 | K, Si, Se, S |
| BD 503 | N, Ca, Mn, B |
| BD 504 | Fe, K, Ca, Mg, S |
| BD 505 | Ca, P |
| BD 506 | K, Si, Mg, B, Se |
| BD 507 | P, Se, Mg |

In addition, the plastering of the compost heap aids in developing heat within the heap and also prevents the volatilisation of its nutrients. The heat produced in the compost heap by bacteria generally kills the seeds of weeds and rids it of disease-causing microbes. It has been found in India that the use of these biodynamic preparations (BD502 to BD507) hastens the composting process. In addition to biodynamic preparations, different microbial cultures like Trichoderma sp. Pleurotus sp. and EM (effective micro-organisms) solutions are used to activate the composting process.

## INPUTS FOR PEST MANAGEMENT

### Cultural Methods

A 'pest' is any organism that is responsible for economically significant production losses in agriculture and the term may include insects, microorganisms, weeds, rodents, etc. According to an estimate by the WHO, about one million cases of pesticide poisoning with 20,000 deaths occur every year due to pesticide residues in food including cereals, pulses, vegetables, fruits, oils, milk and milk products, fish, poultry, meat and water. Pests continue to exert a heavy toll in terms of yield losses. The last global estimate of yield losses due to crop diseases, insects and weeds for eight major crops including rice, wheat, maize, cotton, soybean, barley and coffee indicated these were equivalent to approximately US$ 250 billion in value annually.

The age-old methods of 'catch and kill' for insect pests and 'cut and burn' for diseases are still being practised by farmers on several crops. Various mechanical methods like the bamboo bow trap for the control of rats; the erection of owl perches in rice fields for rodent control; the provision of tin sheets around coconut trunks to prevent rats from damaging the nuts are still widely adopted. The use of botanicals such as neem is quite popular. Biological control using the natural enemies of insect pests e.g., the introduction of predatory red ant colonies into orchards or encouraging bird predators by installing perches are still techniques in use. Monitoring pest and disease occurrence at regular intervals is very essential for effectively managing both.

### Insect Pest Management

Different approaches have been adopted to manage insect pests. Since millennia in India, farmers have been modulating crop growth through adjustments in the time of sowing, cultural practices, and by controlling plant population density for altering the microclimate to minimise risks. These cultural practices alter the epidemic pyramid, namely, the pest/pathogen, host environment and time.

#### *Crop rotation*

Pest problems can be reduced to a large extent by carefully rotating a crop belonging to one family with another from a different group. Rotating groundnut with maize will reduce white grub attacks. Rotating pigeon pea or chickpea with a non-leguminous crop is a good practice for managing fusarium wilt and nematode problems.

*Trap crops*

Raising trap crops as inter and/or border crops is an important cropping system approach. The trap crop must be distinctly attractive to the pest when compared with the main crop. It provides protection either by preventing the pest from reaching the main crop or by restricting them to a certain part of the field where they can be economically destroyed. Mustard along with cabbage is a trap crop for the control of the diamond back moth, aphids and the leaf webber. African marigold is a good trap crop for the American boll worm; besides it also attracts the adults of the leaf minor to lay eggs on its leaves. Maize plants can be a trap crop to attract fruit fly adults in vegetable cultivation where the fruit fly is a major problem.

*Intercropping*

Intercropping with diverse plant geometry breaks the pattern of standard mono-cropping and limits infestation from insects. Changing the method of planting in certain crops by adopting a 'paired row system' provides the possibility of introducing short duration intercrops without affecting the plant population of the main crop which can be either a medium or long duration crop like cotton or sugarcane. Diverse planting not only obstructs adults from egg-laying but also the release of volatile allelo-chemicals from particular crops. These effects do not provide suitable conditions for the development of a microclimate favouring the multiplication of a particular type of pest. For example cabbage, along with carrot or tomato, is an important intercrop combination to effectively manage diamond back moth.

*Use of resistant/t olerant varieties*

Selection of resistant or less susceptible varieties is a very important factor when examining the diversity and intensity of pests in a particular place. Host plant resistance provides a built-in ability to ward off pests. A series of resistant varieties have been developed for all crops in accordance with their agro-climatic zones.

Biological approaches to pest management comprise the use of:

— plants or botanicals
— microbial pesticides
— biocontrol by insects
— biorationals

*Botanicals*

The plant kingdom is a rich storehouse of biologically active compounds and various plant products have been in use for many centuries in India to minimise losses in crops and grain storage. Grainage et. al., compiled a database of plant species that possess insecticidal, antifeedant, repellant, attractant and growth inhibiting properties.

Some of the plants widely used in the preparation of botanical pesticides are

a) *Anona sp*
b) *Asadirachta indica*
c) *Chrysanthemum sp.*
d) *Cymbopogan sp.*
e) *Nicotiana sp*
f) *Pongamia sp*
g) *Vitex sp.*

The seeds, leaves, extracts, fruits, kernels, oil and decoctions of botanicals are used to manage pests.

### *Microbial pesticides or biopesticides*

The use of microorganisms as biocontrol agents is gaining importance in recent years. Biopesticides are living organisms or parts derived from them which are used as control agents to protect the crops from insect pests. Entomopathogenic viruses of the baculovirus group, bacterial insecticides, particularly Bacillus thuringiensis, entomo-fungal pathogens, protozoans and insect parasitic nematodes have been found to control important pests of crops.

### *Types of microbial biopesticides*

#### *Bacterial biopesticides*

The most important bacterial biopesticide is Bacillus thuringiensis, a pathogen of lepidopteran pests infecting crops like cotton, vegetables, etc. Pseudomonas fluorescens, another most commonly used bacterial biopesticide, is effective against wilts and root rot diseases.

#### *Fungal biopesticides*

Fungal biopesticides are divided into two categories, i.e., entomopathogenic fungi and antagonistic fungi. Entomopathogenic fungi infect insects whereas antagonistic fungi attack disease causing fungi and other parasitic microorganisms on the plant. Examples of entomopathogenic fungi include *Beauveria bassiana*. *Verticillium lecanii* and *Metarhizium anisopliae*; antagonistic fungi include *Trichoderma sp*. There are certain fungi which trap nematodes.

#### *Biocontrol by insects*

Here insects are used to control insects. Beneficial insects either prey on insects that have become pests or damage the different stages of insect development like egg, larva, pupa. Accordingly, these biocontrol agents are categorised as:

*Egg parasites:*These parasites damage the egg stage of the harmful insect. Some of the commonly used egg parasites are Trichogramma spp; Telenomus spp; and Testrastichus sp. These parasites control the top shoot borer in sugarcane, the internode borer also in sugarcane, cotton bollworms, paddy stem borers, sorghum stem borers, fruit borers, etc.

*Larval parasites:* These parasites destroy the larval stages of insect pests. A classical example is the use of Bracon spp. for controlling black-headed caterpillar in coconut and Goppniozus nephantidis against the coconut leaf eating caterpillar.

*Pupal parasites*: The pupal stage of insect pests is destroyed by pupal parasites. Testrastichus sp is widely used to control pests like American boll worm, paddy leaf rollers, black-headed caterpillars, etc., in their pupal stages.

*Predators*: Predators like Chrysopa sp, Menochilus spp. are very useful in controlling a wide variety of pests including aphids, whiteflies, cotton boll worm, leaf insects, etc.

***Biorationals***

Biorationals comprise basically of the use of pheromones. Pheromones are chemicals produced by insects used in communication. They are exploited by insects for feeding, mating, following a trail, triggering alarms, etc. Insect sex attractants are also pheromones produced by one sex to attract the opposite sex for the purpose of mating. They are species specific and are produced by females to attract males with few exceptions. Their potency and selectivity offer considerable potential for the manipulation of pests. Many sex attractants have now been synthesized and are available in commercial forms. There are three principal ways in which they may be used to assist the control of pests:

— through traps;
— by disrupting the mating process;
— for survey and monitoring.

## Crop Disease Management

Disease management involves three components: pathogen, host and environment. Pathogen management involves reduction in inoculums and eradication and prevention of inoculums. Management of the host involves improving plant vigour and inducing resistance through nutrition and breeding. The environment involves water soil and crop management, i.e., etiology of the insect. Different approaches adopted in these areas are discussed below:

### *Cultural practices*

Cultural practices are a very important component of crop disease management and directly related to soil microbial biodiversity, crop health and sustainable agriculture. Crop rotation with cereals is quite effective in reducing incidence of root knot nematode and pea wilt. Fallowing is effective against Sclerotina sp and Verticillium sp as these pathogens quickly degenerate under the influence of alternate drying and wetting of soil. Mixed cropping and large scale cropping of multiple genotypic lines are better than monoculture of genotypes to avoid epiphytotics of a disease. Plant spacing adjustment should be done to avoid development of congenial microclimates.

### *Use of resistant varieties*

Growing lines/varieties that are resistant/tolerant to a particular disease prevalent in the area is the best method of disease management.

*Use of biocontrol agents*

Several antagonists have been isolated and many of them are being utilised in the management of soil borne pathogens. Some of the antagonists are highly specific to a particular pathogen but most fungal antagonists have a broad spectrum host range on which they parasitise.

The following biocontrol agents belonging to different groups of microorganisms have been identified:

— Fungi – *Trichoderma sp., Aspergillus sp., Penicillium, Verticillium sp.*

— Bacteria – *Pseudomonas spp., Bacillus spp.*

— Actinomycetes – *Streptomyces spp.*

Soil application and seed treatment are the only feasible methods of using bio-agents under field conditions.

The optimal utilisation of any biocontrol agent in soil is only possible when it is introduced with well-decomposed organic amendments, congenial moisture and protected from extreme weather conditions till it is established in the soil.

Biological seed treatment can be done by priming of seed, soaking seed, seedling dipping, dry powder treatments, etc., depending upon the nature of the biocontrol agents. Generally 5 gm per kg of seed on a commercial formulation basis is used for seed treatment but spore concentration should be between 106 to 109/ml. Similarly, 10–25 gm of the formulation should be applied in a square metre area depending upon the soil texture.

**Weeds**

Weeds are defined as unwanted plants and are also considered as pests. There is a need to redefine the concept of weeds as plants whose use is not yet known to human beings. Weediness is in the eye of the beholder. Weeds are nature's teachers and are good bio-indicators of soil health.

As the fertility of the soil declines, the land is invaded by noxious and virulent weeds and grasses. Only by following sound husbandry practices can we keep these invaders at bay.

In fact, many weeds have long, probing root systems that help break up the subsoil in order to mine the mineral wealth from the depths. This is brought up to the surface and added to the top few inches of the ground when the foliage dies.

Weed roots also improve the structure of soil. Fine delicate feeder roots bind sand particles that would otherwise be blown away. The movement of the roots through the soil breaks open clays. By covering the surface of the ground, vegetation prevents soil erosion and slows the deterioration of bare soils and thereby, the leaching away of plant nutrients. When weeds die, they contribute vast amounts of organic matter to the ground. One acre of roots adds as much as three tonnes of compost. Weeds can be kept under check by adopting the various approaches associated with organic agriculture discussed below:

— Certain weed species like barnyard grass are often associated with particular crops like rice and wild oat; or Phalaris minor in wheat, dodder in alfalfa, etc. Populations of such weeds usually will increase whenever the crop is grown on the same field continuously for several years.

— Crop rotation helps in different ways to manipulate and reduce the weed population.

— Mulching the soil surface can prevent weed seed germination or physically suppress seedling emergence. In cereals, an understorey of clover (*Trifolium spp*) has been shown to improve soil fertility and suppress weeds. The use of Azolla in paddy fields suppresses the emergence of annual weeds.

— Planting time considerably influences the occurrence and manifestation of weed species. Delayed sowing of lentils and chickpea will reduce the infestation by orobanche, a root parasite. However, in most of the winter pulses and in dryland agriculture delayed sowing may be a risky preposition.

— Planting density and pattern will modify crop canopy structure and in turn influence weed smothering ability. Narrow row spacing will bring about variations in the microclimate viz., light intensity, evaporation and temperature at soil surface. The increased shading of soil surface will smother weed growth.

— The method of irrigation also influences weed density. Maintaining flooded conditions in rice fields prevents weed germination, while in drip irrigation, weed density is greatly reduced.

— Biological control of weeds has been attempted with little success. With appropriate technology, weeds can be efficiently utilised for making compost, plant tonics and bio-pesticides all of which have a great potential in organic agriculture.

## REFERENCES

Dikshit, P. K., Report on 'Organic Farming' prepared by a Technical Committee constituted by Ministry of Agriculture, Govt. of India. (Convenor: P. K. Dikshit), 1993.

Douglas, C., Gerard D'Souza and T. Phipps, 'Alternative agriculture adoption: Effect of ground water contam mination and other factors', R.M publications No 91, 1991.

# 4

# Relevance of Organic Farming

Organic agriculture is one among the broad spectrum of production methods that are supportive of the environment. Organic production systems are based on specific standards precisely formulated for food production and aim at achieving agro ecosystems, which are socially and ecologically sustainable. It is based on minimising the use of external inputs through use of on-farm resources efficiently compared to industrial agriculture. Thus the use of synthetic fertilisers and pesticides is avoided.

'Organic' in organic agriculture is a labelling term that denotes products that have been produced in accordance with certain standards during food production, handling, processing and marketing stages, and certified by a duly constituted certification body or authority. The organic label is therefore a process claim rather than a product claim. It should not necessarily be interpreted to mean that the foods produced are healthier, safer or all natural. It simply means that the products follow the defined standard of production and handling, although surveys indicate that consumers consider the organic label as an indication of purity and careful handling. Organic standard will not exempt producers and processors from compliance with general regularity requirements such as food safety regulations, pesticide registrations, general food and nutrition labelling rules, etc.

Ethical issues such as fair labour practices and animal ethics have also been included in organic agriculture definitions. All these however primarily focus on ecological principles as the basis for crop production and animal husbandry. To promote organic agriculture and to ensure fair practices in international trade of organic food, the Codex Alimentarius Commission, a joint body of FAO/ WHO framed certain guidelines for the production, processing, labelling and marketing of organically produced foods, with a view to facilitate trade and prevent misleading claims.

The Codex Alimentarius Commission defines organic agriculture as a holistic food production management system, which promotes and enhances agro ecosystem health, including biodiversity, biological cycles and soil biological activity. It emphasizes the use of management practices in preference to the use of off-farm inputs, taking into account that regional conditions require locally adapted systems. This is accomplished by using, where possible, agronomic, biological and mechanical methods, as opposed to using synthetic materials, to fulfil any specific function within the system.

### ORGANIC AND CONVENTIONAL AGRICULTURE

The issues on organic agriculture fall under the following six categories:

(i) Can organic farming produce enough food for everybody?

(ii) Is it possible to meet the nutrient requirements of crops entirely from organic sources?

(iii) Are there any significant environmental benefits of organic farming?

(iv) Is the food produced by organic farming superior in quality?

(v) Is organic agriculture economically feasible? (vi) Is it possible to manage pests and diseases in organic farming?

## Growth of Food Production

### *Organic agriculture and yield*

Yields relative to comparable conventional systems are directly related to the intensity of farming of the prevailing conventional systems. This is not only the case for comparison between regions, but also between crops within a region, and for individual crops over time. An over-simplification of the impact of conversion to organic agriculture on yield indicates that:

(i) In intensive farming systems, organic agriculture decreases yield; the range depends on the intensity of external input use before conversion.

(ii) In the so-called green revolution areas (irrigated lands), conversion to organic agriculture usually leads to almost identical yields.

(iii) In traditional rain-fed agriculture (with low external inputs), organic agriculture has shown the potential to increase yields.

A number of studies have shown that under drought conditions, crops in organic agriculture systems produce significantly higher yields than comparable conventional agricultural crops, often out-yielding conventional cropsby 7–90%. Others have shown that organic systems have less long-term yield variability. A survey of 208 projects in developing tropical countries in which contemporary organic practices were introduced, showed average yield increases of 5–10% in irrigated crops and 50–100% in rainfed crops.

The so-called organic transition effect, in which a yield decline in the first 1–4 years of transition to organic agriculture, followed by a yield increase when soils have developed adequate biological activity, has not been borne out in some reviews of yield comparison studies. Trials conducted on organic cotton at Nagpur indicated that after the third year, the organic plot, which did not receive fertilisers and insecticides, produced as much cotton as that cultivated with them. Similarly, studies conducted in Punjab clearly indicated that organic farming gave higher or equal yields of different cropping systems compared to chemical farming after an initial period of three years.

*Organic agriculture and food security*

The common claim that large-scale conversion to organic agriculture would result in drastic reduction in world food supplies or large increases in conversion of undisturbed lands to agriculture, has not been borne out in modelling studies. Conversion studies showed that domestic food consumption would not suffer, exports would vary depending on crop, but the structure of farming would definitely change with more diversification of agriculture. Widespread conversion to organic agriculture would result in crop yield increase over the current averages as a result of increased investment in research and extension.

In northern Europe domestic food demand could be met with organic methods, but food would be more expensive. Similar conclusions were also reached in studies in the US. A German study concluded that if the per capita dietary calories from meat were reduced to 21% from the current 39%, all German crop lands could be converted to organic agriculture without an increase in imports or expansion of agricultural land. According to the report such a conversion would be possible by 2017, as the survey indicated that 40% of German youth have planned to maintain low or no-meat diets. The threshold for this conversion may come sooner, as abandonment of meat-consumption is accelerating in Europe as a result of BSE. Several studies have proved the benefits of vegetarian diet over non-vegetarian in terms of energy consumed for food production as well as other nutritional advantages.

Organic farmers grow a variety of crops and maintain livestock in order to optimise use of nutrients and the space between species. This ensures economic advantages through low crop production or yield failure due to biotic and abiotic factors in all of these simultaneously. This can have an important impact on local food security and resilience. In rain-fed systems, organic agriculture has demonstrated to out-perform conventional agricultural systems under environmental stress conditions. Under the right circumstances, the market returns from organic agriculture can potentially contribute to local food security by increasing family income. At the global level, especially in developing countries with high population pressure, and with the present state of knowledge and technology, organic farmers cannot produce enough food for everybody.

*Nutrient management*

Organic farming is often understood as a form of agriculture with use of only organic inputs for the supply of nutrients and management of pests and diseases. In fact, it is a specialised form of diversified agriculture, wherein problems of farming are managed using local resources alone. The term organic does not explicitly mean the type of inputs used; rather it refers to the concept of farm as an organism. Often, organic agriculture has been criticised on the grounds that with organic inputs alone, farm productivity and profitability might not be improved because the availability of organic sources is highly restricted. True, organic resources availability is limited; but under conditions of soil constraints and climate beggaries, organic inputs use has proved more profitable compared to agrochemicals.

Organic farming systems rely on the management of soil organic matter to enhance the chemical, biological and physical properties of the soil. One of the basic principles of soil fertility management in organic systems is that plant nutrition depends on 'biologically-derived nutrients' instead of using readily soluble forms of nutrients; less available forms of nutrients such as those in bulky organic materials are used. This requires release of nutrients to the plant via the activity of soil microbes and soil animals. Improved soil biological activity is also known to play a key role in suppressing weeds, pests and diseases.

Animal dung, crop residues, green manure, biofertilisers and bio-solids from agro-industries and food processing wastes are some of the potential sources of nutrients of organic farming. While animal dung has competitive uses as fuel, it is extensively used in the form of farmyard manure. Development of several compost production technologies like vermicomposting, phosphocomposting, N-enriched phosphocomposting, etc. improves the quality of composts through enrichment with nutrient-bearing minerals and other additives. These manures have the capacity to fulfil nutrient demand of crops adequately and promote the activity of beneficial macro- and micro-flora in the soil.

There are several doubts in the minds of not only farmers, but also scientists about whether it is possible to supply the minimum required nutrients to crops through organic sources alone. Even if it is possible, how are we going to mobilise the organic matter? At this juncture, it is neither advisable nor feasible to recommend the switchover from fertiliser use to organic manure under all agro-ecosystems. Presently, only 30% of our total cultivable areas has irrigation facilities where agrochemicals use is higher compared to rain-fed zones. It is here that ingenuity and efforts are required to increase crop productivity and farm production despite recurrence of environmental constraints of drought and water scarcity.

The basic requirement in organic farming is to increase input use efficiency at each step of the farm operations. This is achieved partly through reducing losses and adoption of new technologies for enrichment of nutrient content in manure. Technologies to enrich the nutrient supply potential from manure, including farmyard manure three to four times are being widely used in organic farms. According to a conservative estimate, around 600 to 700 million tonnes (mt) of agricultural waste is available in the country every year, but most of it is not used properly. We must convert our filth into wealth by mobilising all the biomass in the rural and urban areas into bioenergy to supply required nutrients to our starved soil and fuel to farmers. India produces about 1800 mt of animal dung per annum. Even if two-thirds of the dung is used for biogas generation, it is expected to yield biogas not less than 120 m $m^3$ per day. In addition, the manure produced would be about 440 mt per year, which is equivalent to 2.90 mt N, 2.75 mt $P_2O_5$ and 1.89 mt $K_2O$.

Organic farms and food production systems are quite distinct from conventional farms in terms of nutrient management strategies. Organic systems adopt management options with the primary aim to develop whole farms, like a living organism with balanced growth, in both crops and livestock holding. Thus nutrient cycle is closed as far as possible. Only nutrients in the form of food are exported out of the farm. Crop residues burning is prohibited; so also the unscientific storage of animal wastes and its application in the fields. It is, therefore, considered more environment friendly and sustainable than the conventional system. Farm conversion from high-input, chemical-based system to organic system is designed after undertaking a constraint analysis for the farm with the primary aim to take advantage of local conditions and their interactions with farm activities, climate, soil and environment, so as to achieve (as far as possible) closed nutrient cycles with less dependence on off-farm inputs. This implies that the only nutrients leaving the farm unit are those for human consumption.

Crop rotations and varieties are selected to suit local conditions having the potential to sufficiently balance the nitrogen demand of crops. Requirements for phosphorus, sulphur and micronutrients are met with local, preferably renewable resources. Organic agriculture is, therefore, often termed as knowledge-based rather than input-based agriculture. Furthermore, organic farms aim to optimise the crop productivity under a given set of farm conditions. This is in contrast to concept of yield maximisation through the intensive use of agrochemicals, irrigation water and other off-farm inputs. There are ample evidences to show that agrochemical-based, high-input agriculture is not sustainable for long periods due to gradual decline in factor productivity, with adverse impact on soil health and quality.

### *Environmental benefits*

The impact of organic agriculture on natural resources favours interactions within the agro-ecosystem that are vital for both agricultural production and nature conservation. Ecological services derived include soil forming and conditioning, soil stabilisation, waste recycling, carbon sequestration, nutrient cycling, predation, pollination and habitats. The environmental costs of conventional agriculture are substantial, and the evidence for significant environmental amelioration via conversion to organic agriculture is overwhelming. A review of over 300 published reportsshowed that out of 18 environmental impact indicators (floral diversity, faunal diversity, habitat diversity, landscape, soil organic matter, soil biological activity, soil structure, soil erosion, nitrate leaching, pesticide residues, $CO_2$, $N_2O$, $CH_4$, $NH_3$, nutrient use, water use and energy use), organic farming systems performed significantly better in 12 and performed worse in none. There are also high preconsumer human health costs to conventional agriculture, particularly in the use of pesticides. It is estimated that 25 million agricultural workers in developing countries are poisoned each year by pesticides.

### *Safety and quality of food*

There is a growing demand for organic foods driven primarily by the consumer's perceptions of the quality and safety of these foods and to the positive environmental impact of organic agriculture practices. The 'organic' label is not a health claim, it is a process claim. It has been demonstrated that organically produced foods have lower levels of pesticides and veterinary drug residues and in many cases lower nitrate contents. No clear trends have, however, been established in terms of organoleptic quality differences between organically and conventionally grown foods.

There have been many claims that eating organic foods increases exposure to microbiological contaminants. But studies investigating these claims have no evidence to support them. Organic foods must meet the same quality and safety standards applied to conventional foods. These include the CODEX General Principles of Food Hygiene and Food Safety Programmes based on the Hazard Analysis and Critical Control Point. Analysis of pesticide residues in produce in the US and Europe has shown organic products have significantly lower pesticide residues than conventional products. Nitrates are significant contaminants of foods, generally associated with intensive use of nitrogen fertilisers. Studies that compared nitrate contents of organic and conventional products found significantly higher nitrates in conventional products.

There are also claims that food produced by organic methods tastes better and contains a better balance of vitamins and minerals than conventionally grown food. However, there is no clear scientific evidence, with some studies showing increase in vitamin C, minerals and proteins, more sweeter and less tart applesand others not. A crude analysis of the literature, however, favours organic products in this area. A tasting panel convened by the Consumer Association in the United Kingdom did not consistently favour the taste of organic fruits and vegetables. Quality after storage has been reported to be better in organic products relative to conventional products after comparative tests. Reviews of organic vs conventional product sensory analysis studies have reported results that do not clearly substantiate claims of superior organic product tastiness.

It is a known fact that the quality of crops is controlled by a complex interaction of factors, including soil type and the ratio of minerals in added compost, manure and fertiliser. So it is difficult to separate the influence of the environment and farming system.

*Economics of organic farming*

The replacement of external inputs by farm-derived resources normally leads to a reduction in variable input costs under organic management. Expenditure on fertilisers and sprays is substantially lower than in conventional systems in almost all the cases. In a few cases, higher input costs due to the purchase of compost and other organic manure have been reported. Studies have shown that the common organic agricultural combination of lower input costs and favourable price premiums can offset reduced yields and make organic farms equally and often more profitable than conventional farms. Studies that did not include organic price premiums have given mixed results on profitability.

Studies from Europe and Canada show labour costs in organic agriculture average 40–50% higher where the wage rates are generally higher. Gross margins, the difference between farm output and variable costs are generally similar or, where there are favourable price premiums, higher in organic agriculture. The economics of organic cotton cultivation over a period of six years indicated that there is a reduction in cost of cultivation and increased gross and net returns compared to conventional cotton cultivation in India.

### *Pest and disease management*

Pest control in organic farming begins by making sensible choices, such as growing crops that are naturally resistant to diseases and pests, or choosing sowing times that prevent pest and disease outbreaks. Careful management in both time and space of planting not only prevents pests, but also increases population of natural predators that can contribute to the control of insects, diseases and weeds. Other methods generally employed for the management of pests and diseases are: improving soil health to resist soil pathogens and promote plant growth; rotating crops; encouraging natural biological agents for control of diseases, insects and weeds; using physical barriers for protection from insects, birds and animals; modifying habitat to encourage pollinators and natural enemies of pests; and using semi-chemicals such as pheromone attractants and trap pests.

Organic farmers have long maintained that synthetic fertilisers and pesticides increase crop susceptibility to pests. Research substantiates some of these claims. Organic crops have been shown to be more tolerant as well as resistant to insect attack. Organic rice is reported to have thicker cell walls and lower levels of free amino acid than conventional rice. Plant susceptibility to insect herbivory has been shown in numerous studies to be associated with high plant N levels related to high inputs of soluble N fertilisers. Free amino acids, associated with high N applications, have been reported to increase pest attack.

Soil-borne root diseases are generally less severe on organic farms than conventional farms, while there were no consistent differences in foliar diseases between the systems. The successful control of root diseases in organic systems is likely to be related to the use of long and diverse crop rotations, crop mixtures and regular application of organic amendments. Increased levels of soil microbial activity leading to increased competition and antagonism in the rhizosphere, the presence of beneficial root-colonising bacterial and increased levels of vesicular-arbuscular mycorrhisal colonisation of roots have all been identified as contributing factors in the control of root diseases.

## REFERENCES

International Federation of Organic Agriculture Movements, IFOAM, Basic Standards for Organic Production and Processing, IFOAM Publications, Tholey-Tholey, Germany, 1998.

Kortbech-Olesen, R., "Export opportunities of organic food from developing countries", In *World Organics,* Agra Europa (London) Ltd, London, UK, 9–10 May 2000.

Mohan Singh, "Organic farming prospects in Indian agriculture", *Souvenir of 68th Annual Convention of Indian Society Soil Science,* CSAU&T, Kanpur, 4–8 November 2003.

Padel, S. and Lampkin, N. H., *The Economics of Organic Farming: An International Perspective,* CAB International, Wallingford, UK, 1994.

# 5

# Sustainable Organic Plant Breeding

Organic producers try to achieve an optimal economic yield without exhausting natural resources in and around the production site while adhering to organic criteria. Instead of synthetic chemical fertilisers, veterinary drugs, pesticides, herbicides and growth hormones, more natural principles and methods are applied. Organic farmers strive for closed production cycles, natural self-regulation and agro-biodiversity.

Organic farming is more than just an environmentally-friendly production method, it is also a source of wholesome food. When asked why they believe that organic produce is better for them, consumers most often refer to the fact that organic crops are not treated with chemical pesticides, i.e. that they are free of residues. However, we want to develop a vision of food quality that encompasses more than just the absence of residues.

Plant breeding systems specifically for organic production are few and far between, even though the organic sector is striving to achieve a completely organic production chain. Until now, organic farmers have often had little choice but to use conventionally bred varieties designed for conventional systems based on the use of synthetic chemical fertilisers and pesticides. Despite these varieties' higher yield compared to old varieties—on organic farms, too—performance in organic systems could be vastly improved if varieties were bred for typically organic conditions.

The organic sector has listed its wishes regarding plant characteristics (product quality) and methods to develop new varieties (production quality). These methods should take into account organic principles about breeding techniques as well as related socio-economic conditions and structures.

## PARTIES IN ORGANIC PLANT BREEDING

Numerous people are involved in the production chain from breeder to consumer. There are three main parties: farmers (growers and producers), trade and industry (manufacturers, retailers and wholesalers) and breeders (gene banks, seed companies, etc.). The relationships between these domains are pictured in figure 1.

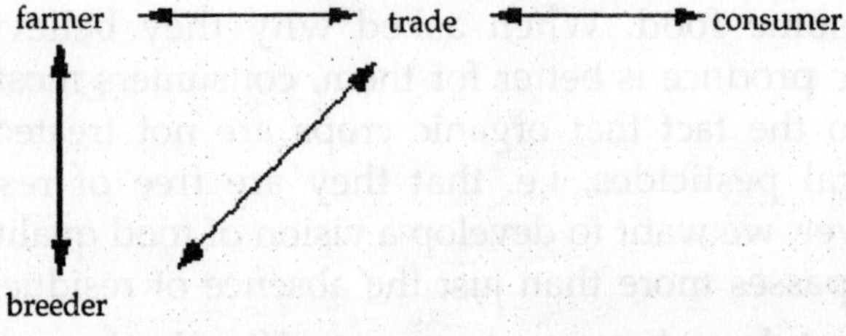


**Figure 1. The three main parties in the organic plant breeding chain**

### Farmers

An organic farmer is dependent on organic fertilisers and nitrogen-fixing legumes for mineral inputs and this in turn makes him dependent on regional and local circumstances. From an organic point of view, the farm is a functioning organism, a natural element in a dynamic environment.

Improving performance, therefore, is a matter of repeated fine-tuning and precision regulation. In this process, organic farmers gain knowledge of and experience in dealing with regional and local possibilities and limitations. This experience explains their preference for varieties that perform well in many different growing environments and that are resilient enough to be able to adapt independently to natural seasonal variations in temperature and mineralisation. Such varieties result from breeding and selecting in specific regional and organic field situations.

### Trade and Industry

Parties in trade and industry, including consumers, demand products that meet certain quality standards (product grade, keeping quality, sorting quality and size). Consumers expect organic products to be good for their health and to be produced in an environmentally-friendly manner. For that reason, they prefer fresh, preferably unprocessed, tasty produce that keeps well and is free of chemical or synthetic additives. Some consumers also value produce which has a clear link with the season, a particular region or producer. Partly in response to consumers' wishes, organic retailers are increasingly demanding varieties with species-specific and regional qualities, for example in taste, bouquet, colour and keeping quality.

### Breeders

Breeding has become highly specialised, requiring specific experience with and knowledge of crossing and selection methods. Organic breeders should maintain and develop varieties in a sustainable manner, on the principle that these varieties are part of our cultural heritage. In practice, this means maintaining or if necessary increasing genetic diversity. Important conditions for the maintenance of genetic diversity are the free use and exchange of varieties among breeders, and the varieties' ability to reproduce naturally in organic growing conditions.

### ORGANIC PRODUCTION CHAIN

The most important conviction that unites the three links in the organic production chain is that products, production methods and processing methods should respect the wholeness of the living entity as much as possible, to ensure that living entities—the farm organism, the plant organism and the human organism—retain their self-regulating ability.

That is why organic farmers use organic fertilisers, natural pesticides and other natural substances and methods. Why consumers buy fresh, unprocessed produce without chemical or synthetic additives whenever they can. And why organic breeders should only work at the level of the crop and plant. DNA techniques, and to a large extent cell techniques too, occur outside the context of the whole plant in its relation with the environment. These techniques manipulate bits of plants in sealed, artificial laboratory conditions. In view of the goal of organic breeding, they are an ecological detour.

Each party in the breeding process contributes specific knowledge and wishes. Close collaboration between the parties is a must if they are to overcome possible conflicts of interests and agree on a set of breeding goals. In particular, supermarkets' demands regarding product characteristics, which in effect are demands on breeding, can conflict with what organic farmers (or consumers) want. For example, supermarkets only want to sell carrots that are 10-15 cm long, have a uniform shape and a smooth skin. Such carrots can only be produced using hybrid varieties that are lacking in taste.

Organic farmers, too, may have wishes that differ from what conventional breeders are used to. For example, high yielding Golden Delicious is a popular cross-parent in apple breeding (Elstar and Jonagold result from crosses with Golden Delicious). But Golden Delicious is highly susceptible to disease, an undesirable characteristic from an organic point of view. Organic apple breeding programmes, therefore, should not be based on crosses with Golden Delicious.

An interactive approach to breeding may provide that intensity of collaboration which is so crucial to organic agriculture. A good start to interaction is to inform partners in the chain about various processes; for example farmers and trade and industry could learn about organic breeding projects at open houses or during site visits. Breeders should be able to justify their actions to downstream parties, because their products serve the rest of the chain, after all. Variety groups and selection on working organic farms could also play an important role in interactive breeding.

## FARM LEVEL CRITERIA

Most principles of organic agriculture pertain to farm management. Closed production cycles, natural self-regulation and agro-biodiversity are the foundations of a stable agro-ecosystem, which forms the basis of organic production because it achieves the best balance of sustainability and economic feasibility.

### Closed Production Cycles

Most farms that convert to organic production in the Netherlands are specialised farms that depend on mineral inputs (arable production) or fodder and straw inputs (livestock production). In the future, however, organic farms will have to have closed nutrient cycles to reduce losses to an absolute minimum. This can be realised by reuniting arable and livestock production, either by establishing new, mixed farms or by intensive collaboration between specialised farms according to the Partner Farm concept. Such closed farm systems raise new challenges for plant breeders. For example, in closed mixed systems fodder crops must be grown that fulfil both the energy requirement of typical mixed-system livestock and the straw requirement for slurry composting.

### Self-regulation

Soil fertility plays a key role in a self-regulating agro-ecosystem. Various aspects of soil fertility (soil life, humus content, structure) form the basis of healthy crop growth and as such are at the heart of organic farm management. An effective regime of fertiliser applications, biological nitrogen fixation and a carefully planned crop rotation scheme are requisites for good soil fertility. The decision to grow a certain crop is based on market demand and on-farm conditions; care should be taken at all times to prevent depletion of the soil. Compared to conventional farms, organic farming conditions are more diverse, which means that there is at least as much diversity in farm management styles. These different styles of farm management all require their own varieties. This is a task for plant breeders.

### Variety of Organisms: Agro-biodiversity

The greater the diversity of crops and other organisms on a farm, the greater its natural self-regulating capacity. Natural features such as ditch bank vegetation, hedges and flowers provide shelter and food for the natural enemies of parasites. A rich varied soil life, legumes and cover crops contribute to soil fertility and prevent plagues of soil pathogens. Finally, many insects give farmers a helping hand with crop pollination.

### Criteria at Plant Level

When the organic criteria at farm level, above, are extrapolated to the level of the plant, there are three criteria that can be used to in organic breeding programmes:

natural reproductive ability, ability to adapt independently to the environment and genetic diversity while respecting natural species authenticity and species characteristics.

*Natural reproductive ability*

One type of closed production cycle at plant level is the cycle from seed to plant to seed, which we have called natural reproductive ability. We can give several reasons for the preference for varieties that are naturally fertile in organic growing conditions, or that at least have the potential to reproduce naturally. First, in an organic plant breeding system, all activities take place in organic growing conditions, from maintaining parent lines to crossing to selection. This automatically precludes any steps—such as fertilisation—being carried out in laboratories, so quite simply plants must be able to reproduce naturally. Secondly, in organic agriculture a plant's reproductive ability is regarded as a sign of vigour. Vigour is a vital quality which is indicative of health and nutritive value. That is why organic farmers prefer to use varieties that can complete their life cycle in organic conditions. Finally, natural reproductive ability ensures sustainable use of the cultivar, as progeny evolves to keep pace with the times and changing conditions.

*Ability to adapt independently to the environment*

According the concept of "plant x environment interaction", the environment (available nutrients, soil structure, moisture content, air temperature, etc.) and changes to that environment during the plant's growth have just as much influence on a plant's appearance (phenotype) as its genes. In other words, a plant (genotype) is continuously interacting with its environment.

Conventional breeders recognise this to some degree as reflected by their breeding programmes, where trial plots or breeding stations are situated in different geographical regions around the world, so that plants can be selected for local climate conditions. In doing so, conventional breeders seek to develop varieties with a fairly stable phenotype. At the same time, however, conventional farming conditions are adapted as much as possible to the needs of the varieties used. Applications of chemical fertilisers and pesticides reduce inter-local differences in soil fertility and resistance to disease and thus increase a variety's area of distribution.

In organic agriculture, on the other hand, environmental diversity is respected and farmers try to optimise their use of natural resources with minimal synthetic, external inputs. Varieties with a high adaptive capacity perform best in these conditions. In addition, organic farmers have only limited possibilities for intervening when things threaten to go wrong during the growing season. That is why varieties for organic farming must be resilient and flexible. This does not mean that adaptation to local conditions is not important in organic agriculture—it is. The challenge for organic agriculture lies in finding the balance between varieties that are adapted to specific, local growing conditions ("specialised") and varieties that can adapt to different conditions, that have high resilience and therefore perform well in many different environments ("generalised"). What organic farmers really want are varieties adapted to local conditions (specialists) with a high resilience (generalists).

Conventional breeders tend to extrapolate a variety's performance ranking from one environment to another, stating that ranking is independent of environment. This does not mean that it is unnecessary to breed plants for specific environments. Consistency in the ranking of an existing variety does not rule out the possibility that a better performance might have been achieved if the variety had been bred for specific local environmental conditions. An understanding of genotype x environment interactions is required to follow this line of reasoning.

### *Mineral uptake efficiency and rootage*

To give an optimal yield on a certain plot, cultivars must be adapted to in situ seasonal fluctuations in nutrient availability. This applies in particular to soil nitrogen. Mineralisation processes are notoriously slow off the bat in early spring, giving rise to nitrogen shortages in the early stages of growth. Cultivars with a passive and active response to such situations are usually the ones that cope best. Active nutrient mobilisation by the plant (Scheller, 1988) is a combination of deeper rootage and a more efficient nitrogen uptake. These two characteristics are important determinants of a plant's suitability for organic growing conditions and are therefore important selection criteria in organic plant breeding.

### *Weed reduction*

Synthetic chemical herbicides are not allowed in organic farming, so crops must have some intrinsic means of countering weed growth. Vigour and competitiveness, particularly in the seedling stage, are important indicators of weed suppressive ability. The structure of a crop, the number, size and position of the leaves, tillering—especially in poorer conditions—and firmness all contribute to a plant's weed suppressive ability and as such are important selection criteria in organic plant breeding.

### *Resistance to disease and plagues*

Organic agriculture stresses the prevention of diseases and plagues. Varieties that perform well in this respect, especially when supported by crop management measures, tend to have a high polygenetic, or all-round, resistance. Absolute monogenetic resistance is not desirable in organic production as this type of resistance is fairly easily broken. As yet, commercial breeders show little interest in developing varieties with polygenetic resistance because of the work and expense involved and because conventional farmers still demand absolute resistance. A plant's structure, its pattern of growth and its heterozygosity all contribute to field tolerance, resilience and restorative ability. These factors are therefore considered important selection criteria. Preferably, breeders should aim for broad resilience by limiting homozygosity to a functional minimum. In other words, some variation within a variety is desirable. In organic agriculture, the use of heterozygous varieties is more appropriate than sowing a mix of homozygous varieties in the field, which only gives a semblance of diversity.

### *Natural Species Authenticity and Species Characteristics*

#### *Natural species authenticity*

Plants have different defences against cross-pollination with other species, for example, the ability to distinguish pollen on the stigma. These mechanisms help to preserve a species' identity. An often heard question in discussions on the use of biotechnology in plant breeding is whether the line of species authenticity may be crossed. This is a difficult question, especially as there is no clear biological definition of what a species is. Biologists generally use the term species when the following two conditions apply to a group of individual organisms:

- the individual organisms have a high degree of similarity;
- the individuals can produce fertile offspring through sexual reproduction.

There is no black-on-white definition for a species and some plant species are so flexible that they defy any attempt at definition. Take, for example, the wild cabbage (Brassica oleracea), which through domestication has been cultivated to many different forms. And at the other end of the spectrum, spontaneous cross-pollination can give rise to new species, too.

The mere fact that spontaneous crosses do occur naturally among some plants is no reason to freely allow such practices in plant breeding. In nature, spontaneous crosses are the exception rather than the rule and they always occur within set, natural limitations. In organic plant breeding, it may be acceptable to cross the line of natural species authenticity provided fertilisation and embryo growth occur naturally on the plant (i.e. there is no need to resort to techniques to incorporate change at cell level) and progeny resulting from the cross are fertile.

*Plant quality*

A species, regardless of whether it is wild or cultivated, has distinct characteristics which set it apart from other species. These characteristics may pertain to a plant's shape, colour, keeping quality and taste. Although taste is a fairly subjective characteristic, taste does give a variety "character". The perception of taste is subjective, but the description of taste may be objective (think of criteria for winetasting, cheese, tea and tobacco). These characteristics are partly hereditary and partly the result of environmental influences. A variety's characteristics may be changed by breeding, sometimes for the worse as in conventional breeding where too strong a focus on quantity has in many cases diminished quality.

*Region-specific characteristics*

If a cultivar is adapted to or used in a particular environment, it may subsequently acquire specific characteristics resulting from the interaction with that environment. Such region-specific characteristics could be used to promote regional products like bread and beer.

*Nutritional value*

Most cultivars are eaten by people, who believe that fruit and vegetables are good for them because of their nutritional value. However, the concept "nutritional value" is not clearly defined. The general assumption is that "healthy crops make for healthy eating". Presumably then, plants that have grown in a well-balanced, steady manner without the use of pesticides and harmful substances (natural toxins, pesticide residues) and which contain sufficient proteins, minerals and vitamins are healthy. Wellbalanced or harmonious growth takes the middle road between pervasion and stagnation. In biodynamic circles in particular, harmony in the plant's gestalt and ability to mature are considered important factors of food quality.

A plant is said to have developed harmoniously if under the given conditions, its development has been optimal in each phase of growth. For example, in wheat crops, the preference goes to varieties with a clear distinction between vegetative and generative development, which is expressed in the plant's overall development and morphology. Species-specific characteristics such as keeping quality, taste, colour and bouquet will be optimally developed if a cultivar is allowed to grow in a harmonious manner. The sum of these quality characteristics is called "maturity". Since only ripe or mature (harvest-ready) crops are healthy to eat, ability to reach maturity could be adopted as a selection criterion in organic breeding. However, further research is required to develop the concept of nutritional value and how it may relate to plant breeding.

### Socio-economic Criteria

In modern plant breeding, there is a strict division between breeders whose task is to produce new varieties and farmers whose task is to use them. A less distinct division with more consideration for farmer-breeder interaction is believed to be more productive in organic agriculture. Another limitation of conventional plant breeding is its focus on standard varieties and strains that do not take into account the great variety of organic farming conditions. These limitations are economic, institutional, technical and conceptual in nature. Economic, because the costs of a breeding programme and the commercial objectives of propagators result in varieties that cover the largest possible geographical range. Institutional, because of the limited possibilities for breeding crops for specific environments. Technical, because techniques are used which conflict with organic agricultural principles. And finally, conceptual because of conventional breeders' product-oriented approach common to technology development which largely ignores alternative uses of new techniques and their products. These limitations must be overcome if breeding programmes are to be developed which better meet the needs of organic farmers. To this end, breeders should seek closer ties with organic farming practice.

### *Farmer-breeder interaction*

An organic breeding system involves selecting varieties in situ, that is on a working farm. The process of adapting to local production conditions is ongoing in organic breeding. Farmers can play an active part in this selection process, but this does require the relationship between farmers and breeders to be redefined. In this new interactive relationship new propagating material and new varieties could be developed. The interactive approach to breeding is quite common in developing countries, where it is known as participatory plant breeding. Organic agriculture does not oppose specialisation and professionalism provided that there is ongoing interaction at a high level of incorporation, as for example when specialised farms form collaborations. If a farmer does not have the skills required to propagate seed and elects to outsource that part of the plant-seed cycle, it is vital that he or she keeps in touch with the propagator and vice versa. Interaction ensures an optimal use of all available knowledge and experience and ultimately leads to a better breeding result.

*Legislation*

The Seeds and Planting Materials Act protects the intellectual rights of ownership, or breeders' rights. Usually, a variety only qualifies for breeders' rights if it is new, distinguishable, homogeneous (uniform) and stable. Organic agriculture, however, requires varieties which are fairly heterogeneous so that they can adapt to different regional soil and climate conditions, which are resilient and have a high resistance to disease and plagues. The goal is functional homozygosity: breeding plants no further on certain characteristics than is necessary for crop growth and mechanical harvesting. As a result, the varieties may not have that degree of uniformity currently demanded for registration as a variety. Strictly speaking, however, unregistered varieties may not be marketed and this is where legislation and the needs of organic agriculture clash.

Patent law is also a potential threat to organic breeding. In principle, farmers have the right to propagate their own seed, which in organic farming is a crucial step in the continuous process of adapting a variety to local production conditions. This right is called farmers' privilege. The law says that farmers' privilege only applies to unpatented varieties or genes. This is cause for concern as the patenting of new varieties or even of specific genes is becoming commonplace in conventional breeding.

Organic agriculture aims to improve the self-regulating capacity of the agro-ecosystem, both in the ecological sense and socio-economically. Amendments to existing legislation to curtail the power of modern capital-intensive forms of breeding are required to stimulate a greater diversity of breeding initiatives. Specifically, the amendments to regulations regarding intellectual property rights, registration criteria, licensing and protection against environmental and health risks.

*Funding structure*

Professional breeders are only prepared to make these investments if they expect to make a reasonable return. These costs do not encourage breeders to make the extra effort to develop many different varieties for specific conditions. Moreover, cultivation and user value studies will have to be expanded to include criteria for organic farm systems, which might raise costs even further. Another additional cost of an interactive organic breeding system is the need for an organisation which deals with the outsourcing of the selection and propagation process to organic farmers. Quite simply, seed or plant sales alone could never cover the entire cost of an organic breeding programme. A new funding structure based on new sources of funding will therefore have to be developed.

*Cultural diversity*

Conventional varieties are increasingly bred for highly conditioned circumstances and large fields. This inevitably means a loss of genetic diversity. Advances in gene technology and technology patenting are only accelerating this trend. This capital-intensive type of breeding is fast becoming the exclusive domain of multi-nationals that can afford to pay patent holders huge sums of money for a licence or even for the patent itself. Corporate mergers are also narrowing the genetic basis of breeding programmes. Mergers cut the number of breeding programmes for a cultivar, and take away breeders' right to freely exchange crosses and seed. Cultural inbreeding and genetic erosion will be the ultimate result.

In addition to these biological arguments, we can also cite moral or ethical objections to the practice of patenting organisms or parts of organisms (genes). Basically, patenting means that characteristics of organisms which have developed over centuries either in the wild or through the efforts of local farming populations, suddenly become private property simply because that person or company found, identified and isolated the characteristics. This is a denial of centuries of tradition and custom. It also conflicts with the international biodiversity treaty, which states that all life forms have their own intrinsic value.

### Breeding Techniques

One of the main principles of the vision on organic plant breeding is that there should always be interaction between plant and environment, and between farmer and breeder. Ideally, breeding programmes should be developed in consultation with partners in the organic production chain and selection should be carried out by farmers in organic growing conditions. This would result in varieties optimally adapted to organic growing conditions with characteristics that would satisfy all partners in the chain.

#### *Techniques at plant and crop level*

Preferably, interactive organic plant breeders should only use techniques at the level of the plant and crop. Parent lines would have to be maintained in organic conditions, just as progeny would be selected and propagated in organic plots. Techniques at the level of the plant or crop allow farmers to carry out the selection process themselves on their own farm or in their own region, though possibly in consultation with a professional breeder. In this way, organic management and regional influences would be incorporated in the selection process.

#### *Hybrids*

When using hybrids, F1 progeny must be maintained by repeatedly crossing the parent lines, since generative propagation of the F1 defeats the purpose of hybridisation.

This end-of-the-line nature of hybrids is why some—chiefly biodynamic—growers and breeders object to their creation or use. Nevertheless, biodynamic farmers are known to use hybrids simply because seed-propagating varieties with comparable yields are not (yet) available or because buyers' demands regarding uniformity and size can only be met by using hybrids. The objection stated above is not shared to the same extent by ecological growers, although they, too, would welcome improvements in seed-propagating varieties as well as a more relaxed attitude on the part of buyers.

Seed-propagating varieties could equal hybrids in uniformity and yield if they were to become the focus of breeding programmes. Hybrid progeny cannot be generatively propagated, which thus effectively keeps farmers from collecting seed themselves. Although this may be an incentive for breeders to invest in hybrids, such varieties do not serve the long-term interests of organic farmers. For this reason, more effort should be put into the development of seed-propagating varieties.

The use of hybrids in organic agriculture is justified in the following situations:

— comparable seed-propagating varieties are not available;
— homozygous lines still have enough vigour to be propagated in organic growing conditions;
— the F1 is fertile and can be used as a cross-parent, i.e. no cms without restorer lines.

### *Techniques at cell level*

Cell techniques such as the culture of embryos and ovaries are used to incorporate characteristics of wild plants in cultivars. Embryos are cultured after pollination and seed formation on the plant because there is a reduced chance that they will grow successfully on the plant. In such cases, the embryos or ovules are isolated from the plant and placed in a culture containing all the necessary nutrients and hormones. These crosses could be created in natural circumstances, but it would be a case of beating the odds, which would require a much greater number of plants.

Techniques at cell level bypass the direct interaction between the plant and its (future) growing environment. During our discussions, it became apparent that breeding techniques at cell level fall in a grey category between the unacceptable DNA techniques and the approved methods of breeding at the level of the plant. From a biological perspective, the cell is the smallest living entity and hence cell techniques could be used without violating the organic principle to work only with living entities. On the other hand, the smallest unit of life encountered by farmers in daily farming practice is the whole plant in relation to its environment. From this perspective, cell techniques are an ecological detour and as such their use in organic breeding continues to be a point of controversy.

In conventional breeding, however, cell techniques have been used since the 1930s. For some crops, such as tomatoes, they have become a crucial step in the maintenance of the cultivar. In effect, cell techniques are synonymous with breeding programmes based on specific, hereditary, usually monogenetic resistance or characteristics. In organic breeding, however, we are not interested in incorporating absolute resistance, but in developing varieties that are better able to cope with existing growing conditions. Such a plant might have deeper or more elaborate rootage to survive plagues of insects or fungi. Organic breeding objectives are based on a different concept of plant health. More research is required to work out the operational details of this concept and to enable the development of a breeding system that is less dependent on cell techniques.

An immediate ban on cell techniques that have been used for decades might do more harm than good for some crops and varieties, and have dramatic economic consequences for farmers. Obviously, therefore, satisfactory alternatives will first have to be developed. A condition for using such varieties in organic agriculture is that they have the potential to reproduce naturally in the organic growing environment. In addition, characteristics which have been incorporated using cell techniques should also be receptive to traditional cross methods.

In order to stimulate the development of organic breeding programmes which are not dependent on cell techniques. Varieties or lines produced with cell techniques and originating from before a certain date may be used as foundation stock. Varieties and lines produced with cell techniques after 1-1-2001 would not be allowed in green mark breeding programmes.

### *Protoplast fusion*

Protoplast fusion goes one step further than the techniques described above. Protoplast fusion is carried out with loose, unconnected cells (without membranes) that are in no way recognisable as plant material. This technique is applied when two species differ so much that a successful cross could never be achieved by natural methods. This is also a major reason for rejecting the technique in organic plant breeding. The consequences of rejecting this technique are not that far-reaching, as it is fairly new in conventional breeding, too. Protoplast fusion is primarily used to transfer cytoplasmic male sterility from one species to another (for example, from radish to cabbage). The technique has been patented, which is another reason to not use it in organic breeding programmes.

### *Cytoplasmic male sterility (cms)*

Organic agriculture rejects the use of techniques that

negatively affect a plant's natural reproductive ability. The sector particularly opposes cms hybrids without restorer genes. There are plenty of alternative seed-propagating varieties or hybrids without cms. Unfortunately, the use of cytoplasmic male sterility in modern hybrids is growing. Although cytoplasmic male sterility occurs naturally in some species, breeders are currently using protoplast fusion to transfer the characteristic to other species as well. These new hybrids are subsequently patented. The absolute cms in cabbage, for example, originates from radish. Cms hybrids differ from non-cms hybrids in a number of ways:

— the advantage of using cms hybrids in practice is that inbred plants (deviants) do not occur in the field as is the case with non-cms hybrids (sometimes more than 10%);

— other breeders cannot acquire the mother line by searching for inbred plants because these do not exist; this gives the breeder who owns the mother line a competitive advantage;

— other breeders cannot incorporate any of a cms hybrid's hereditary characteristics in their own breeding material; the breeder of the cms hybrid has exclusive access to the gene pool, so that he is the only one who can incorporate the selected characteristics in marketable varieties and profit from them. Ultimately, however, cultural inbreeding will occur and this may lead to genetic erosion. These practices therefore go against the grain of organic agriculture.

Imagine a breeder who sees a promising cms hybrid and wants to cross it with his own lines or plants. The cms plant (plant A) is male sterile (does not produce functioning pollen) and can therefore only be used as a mother plant. Crossing plant A with a fertile father (plant B) yields plants that have a 50% share of genes from each parent, including the cytoplasmic male sterility as that is always inherited from the mother. The only thing the breeder can do now is cross the progeny with a fertile plant. This F2 generation will have inherited 25% of plant A's genes, again including the cms characteristic. Eventually, the breeder will get a generation which has 0% of plant A's genes, but these plants will still be male sterile. The only characteristic that can be transferred from a cms cross-parent is the cms characteristic, and if that is what you want, one cross is all it takes. Every other characteristic is lost in the backcross. And because plant A is sterile, there is no way of using it other than as the mother plant.

The free exchange of varieties, hybrid or otherwise, will always form the basis of sustainable breeding. This was one of the primary considerations that led to the establishment of breeders' rights for plants, as an alternative to patenting. If every single variety was patented, breeding would be made impossible. But cytoplasmic male sterility, too, makes breeding impossible.

Breeders who work with cms can use other breeders' non-cms varieties, but they in turn cannot use the cms varieties. There is no mutual exchange of benefits or sharing of profits. This one-way breeding is a form of parasitism. Breeders have a responsibility to maintain cultivars because they are part of our cultural heritage, resulting from centuries of cultivation and breeding. These cultivars cannot be maintained without natural reproduction and genetic transfer.

As cms becomes more and more common in modern varieties, fewer varieties will remain for use in breeding programmes. Eventually, modern varieties will no longer be able to be crossed to broaden the genetic basis of a breeding programme. Programmes will stagnate due to the erosion of genetic diversity. Thus, although cms may seem like an attractive way of getting ahead of the competition, it is ultimately a dead-end road. In effect, each cms hybrid is like a dead twig on a cultivar's family tree.

Breeders will need restorer genes for the infusion of new genetic material. These genes could be transferred from one species to another through genetic engineering. But it is likely that this quality will be patented just as cms was and therefore only be available for cms breeders.

Breeders who do not use cms will always be in an underdog position and on top of that, will end up in a cut-throat competition with each other. If there are no restorer genes, these breeders will not be able to incorporate the genes in cms varieties. At the same time, however, cms breeders will gratefully cash in on each improvement in non-cms varieties. Breeders who use cms techniques are not taking their responsibility to maintain cultivars and in the long term they will drive non-cms breeders into economic extinction. The question is whether, in the current system of breeding, non-cms programmes can feasibly co-exist with cms programmes.

*Artificial induction*

The artificial induction of mutations in DNA by radiation, chemical mutagens or other means is completely random and largely harmful. According to our vision of organic plant breeding, direct manipulation at DNA level is undesirable. Artificial induction is therefore unsuitable for use in organic breeding. An immediate ban on this technique will have little impact, as the technique is currently rarely used in conventional breeding. The same applies to radiated mentor pollen.

*Techniques at DNA level*

Genetic modification is not allowed in organic agriculture. This is justified on the basis of the following organic principles:

— It is a one-dimensional, drastic intervention in a plant's genetic make-up which upsets the plant's relation to the environment. The desired DNA can only be incorporated by first reducing the plant to cell level and then "reconstructing" it using tissue culture techniques. This destroys the historical plant x environment relationship.

— It is a reductionist method that increases the chances of undesirable and unpredictable ecological and health risks in the long term.

— It is a capital-intensive type of breeding in which multinationals hold patents on living organisms and impede a free exchange of genetic material, thus reducing genetic and cultural diversity.

### *DNA diagnostic techniques*

Genetic engineering has also enabled selection at DNA level without actually modifying plants' genetic make-up. These techniques are based on biochemical and molecular markers and may complement field selection methods. The potential of these indirect selection techniques should not be exaggerated, however. Although there is no reason why these techniques should not be applied in organic breeding, selection in the field will always be the primary method of selection because it incorporates the plant x environment interaction.

## MONITORING AND CERTIFICATION

A monitoring system should be set up to enforce regulations governing the use of breeding techniques to ensure that varieties approved for organic farming have indeed been bred and propagated in accordance with organic principles. EU regulations on organic breeding will probably go no further than list approved and prohibited techniques. Most cell techniques, however, leave no trace in plants, so that monitoring for direct evidence of wrongful use would be fruitless. Instead, more indirect monitoring methods will have to be used, for example audits of the production process.

The inspection organisation for organic agriculture in the Netherlands, Skal, carries out process inspections to enforce compliance with EU regulations for organic production. In conventional breeding, too, process inspections are becoming more popular as part of private certification schemes carried out by NAK-AGRO Nederland BV (arable crops), NAK-G (vegetable crops) and NAK-S (ornamental stock). The standards set by private certification schemes may exceed legislative requirements. In addition, such certification systems can be developed and implemented in a relatively short period of time while EU regulations for organic breeding are still pending.

The following categories might be included in varieties lists:

- a black list for varieties that are not allowed in organic agriculture;
- a white list for varieties produced in conventional breeding programmes using approved techniques, including provisionally allowed cell techniques;
- a green list for white-list varieties that have been organically propagated and that perform well in organic growing conditions. These varieties satisfy EU regulations on organic production;
- a green-plus list for varieties that have been produced in organic breeding programmes without cell techniques, which have been organically propagated and which perform well in organic growing conditions.

### Certification System

The following criteria could apply for an organic seed/ plant stock certification system:

- Parent lines must be maintained in organic growing conditions, for example soil cultivation under glass (or mesh).

- Crosses must be made under glass (mesh) or in the field. Parent plants must be rooted in soil (in pots or open soil), not in substrates that do not comply with Skal guidelines. Crosses may only be made using approved crossing techniques.
- Selection of the F1 progeny must be carried out in organic conditions. If necessary, seedlings may be tested under glass, even if the crops are normally grown in the field. However, only approved selection methods may be used. Following generations (F2 onwards) must be selected in normal organic growing conditions. From F3 or F4, selections will be extended to different organic trial plots in different regions. Preferably, these plots should be situated on working organic farms. Farmers participate in the selection process, the outcome of which is subject to environmental influences. In this way, region-specific varieties are developed.
- Varieties must be propagated and maintained in normal organic growing conditions. If desired, these activities may be carried out in soil cultivation under glass or mesh. Only approved propagation techniques may be used.
- The following information should be recorded by farmers in order to apply for certification:
  - the parent lines;
  - the techniques used;
  - selection conditions of the F1;
  - selection conditions of the F2 to Fn;
  - propagation methods and conditions;
  - maintenance methods and conditions.

### Monitoring

Skal could monitor breeders participating in the

certification system, perhaps jointly with NAK. The easiest monitoring system would probably be audits of registration applications. Breeders wanting to register a variety would have to provide detailed information on what techniques they used and whether plants have been selected in organic growing conditions. Breeders would not be required to give a step by step account of the development of the variety.

A breeder could submit a signed declaration that a variety was bred in accordance with organic certification criteria. Skal or NAK could then verify this through an audit of the registration application. If the outcome of the audit is positive, seed/plant stock would be awarded a certificate and/or EKO mark.

### Difference between Organic and Biodynamic Production

Facilitating the development of organic breeding programmes encompasses more than simply laying down a list of approved techniques and an accompanying system for enforcement. Organic breeding is based on a different concept of plant health which requires a new set of regulations and new fundamental or strategic research. The current breeding system does not take into account organic principles. Structural changes will be required before an interactive chain approach, which is crucial to organic breeding, can be realised. A framework for organic breeding cannot be constructed by the organic sector alone; existing (conventional) institutions will have to change their course, too.

### Stimulation Policy

Breeding is a long-term activity and it would be unrealistic to expect the market—which is more focused on the short term—to independently launch a professional organic breeding system when the demand for organic products is still relatively small. Various institutions who participated in the discussions in this project were keen to contribute to an organic breeding system, but do not want to act alone.

Commercial parties, too, showed real interest by actively participating in the discussions. This does not mean that every seed company is ready to make an investment; they will first want to examine the costs and benefits. Naturally, private companies will only want to participate when it is to their benefit and will be unwilling to invest time, money and effort if competitors benefit, too. This applies just as much to breeding companies as it does to organic farmers.

To overcome these difficulties, the government will have to draw up policy to stimulate and finance the development and marketing of organic breeds in a joint effort with private parties. Organisations and institutions need professional skills and money. If they receive no commitment for either, the development of an organic breeding system will surely stagnate.

## FUTURE EXPECTATIONS

Currently, organic agriculture is based on conventional breeding. Additional efforts and investments will be required to set up an organic breeding system for the organic production chain. The conventional farmers—who are also moving towards more sustainable production—will also want to use organic varieties in the future. The project group expects that conventional and organic breeders have more interests in common than currently recognised and that this will become evident in time.

## REFERENCES

Den Nijs, A.P.M. and E. Lammerts van Bueren, "Towards breeding for organic agriculture: cross pollination required between conventional breeders and organic farmers", Prophyta June 1999.

Wolfe, M. and E. Lammerts van Bueren, "A question of breeding", *Organic Farming*, 1999.

# 6

# Organic Transplant Production

Greenhouse grown flower, herb and vegetable transplants create a wide range of production and marketing opportunities.

## STRUCTURE AND EQUIPMENT

Managing the greenhouse atmosphere to provide appropriate light and temperature conditions depends on design, covering materials, heating, cooling and ventilation. For the smallscale farmer, the availability of capital to invest may limit the options.

### Site Selection and Layout

Locate the greenhouse to maximise light interception. At latitudes above 40, the standard recommendation is to orient the greenhouse with the ridge running east and west so that winter light is penetrating the length of the south side of the greenhouse, not from the end. Consider shading from trees and buildings. To avoid shading, locate a distance equal to at least twice the height of the shade source. Buildings attached to greenhouses are located on the north side to prevent shading. the water for irrigation. If there is not adequate slope to move the water quickly away from the greenhouse, provide drainage pipes. The most critical time will be the spring thaw, which is also when the greenhouse will be full and there will be little time to dig drainage ditches.

### Design and Materials

During visits to small-scale diversified vegetable farms and CSA's have seen a wide range of structures used for transplant production. Greenhouse vs cold frame, free standing vs lean-to, heated vs unheated, commercial vs home made; there are many possible options. Decision making depends on the monetary investment possible, available space, and the relative dependence on human verses mechanical environmental regulation.

Most commercial structures will be of galvanised steel pipe or square tubing. Metal end walls and side rails cost more but avoid the issue of treated lumber. The use of wood is influenced by the desire to not use treated lumber, although some certifying agencies differentiate between wood in contact with the ground (not allowed) vs wood used above ground (may be allowed). For hoop houses constructed at MSU, we avoided the use of treated lumber by attaching end wall posts to metal fence posts driven into the ground. We raised the 2"x4" pine baseboards off the ground and attached a readily available 24" x12' fiberglass panel so it went into the ground and provided a wind buffer and something to drop the roll down side against. In other projects where there were more funds available, we purchased and constructed all metal end walls and frames. While manufactured metal endwalls cost 3 or 4 times what wood costs, it looks like a good investment if funds are available.

The most likely covering material will be double layer inflated 3-4 year polyethylene film. Plastic films must be greenhouse grade, not hardware or construction grade. The second most common small farm option is a rigid, single or double wall polycarbonate or acrylic. While glass greenhouses are not likely justified for on farm transplant production, glass windows, sliding patio doors, or glass shower doors can be used for cold frame construction for hardening transplants.

Forced air, natural gas or propane unit heaters are the most common form of heat for small greenhouses producing transplants. One limitation with this design is getting heat down near the plants. Plastic perforated convection tubes can be used to blow warm air beneath benches. Benches with plastic or solid tops and sides can also be located in front of a heater. With cold soil floors it is recommended that flats be separated or lifted off the floor. Bottom heat using hot water circulated.

The shape of the roof will influence light interception. For polyethylene covered structures, the rounder Quonset designs with a flat spot on top are being replaced by a more peaked or gothic design. The higher roof angle increases winter light interception, decreases summer light interception, often increases peak height, helps with snow and condensate removal from the roof, and is highly recommended. Gothic designs come in two pieces which makes shipping easier and less expensive.

**Size**

The primary factors determining size are the desired function, cost/budget, location or space available, method of ventilation and cooling, method of heating and fuel requirement. For the sake of providing some perspective, example sizes relative to small farms and transplant growers might be Asmall@: less than 1000 square feet (20′ x 50′); Amedium@: 2000 up to 4000 square feet (20′ or 30′ x 100′ or more than one); and Alarge@: greater than 5000 square feet. In 1000 square feet at 80% growing space efficiency (a full house), at least 400 flats (11″ x 22″) or 28,800 (72 cell flats) to 80,000 (200 cell flat) transplants can be grown.

## Ventilation

If passive (vents only) ventilation will be used instead of exhaust fans (active ventilation) the greenhouse should be smaller to allow easier cooling. Widths of 20 to 30 feet and lengths of 40 to 50 feet are common. The larger the vent area, the easier it is to cool but usually the harder it is to close up to keep warm when necessary. If roll up sides are used, longer structures can be cooled without fans. Retractable roof designs are also becoming more popular because of the ventilation and cooling possible.

## Heating

beneath the flats is a very efficient way of maintaining growth while keeping air temperatures cool and can be provided with readily available water lines, water heaters and circulating pumps.

## Costs

Complete commercial greenhouses including the structure, utilities, environmental controls, floors, benches, irrigation equipment etc range in cost from as little as $5 per square foot up to $25 per square foot. The upper end might include a glass structure with hot water heat. The cost for the least expensive hoop structure with nothing else added can be between $1.00 and $2.00 per square foot for a 30 x 100 greenhouse that is not assembled. Having something assembled or constructed typically doubles the cost. Shipping costs can also be significant and might be avoided by looking for the most local source. Greenhouse management can be very simple and low tech or very detailed and mechanised. Take the time to learn the key concepts first and save money both in the short and long run. If you are going to build your own greenhouse, take the time to work out how you will integrate light intensity control, ventilation/cooling, and heating. There are choices to be made about the level and cost of mechanisation verses the level of labor input. If you are on the farm (home) most of the time and available to alter vents and shading by hand, the need for automated vents and therefore the cost will be lower. If you don=t want to be tied to the greenhouse, invest in thermostats and ventilation equipment.

When planning a greenhouse for the small farm, it is important to take into consideration the time and labor cost of planning and constructing personal plans or modifications as opposed to paying extra for a complete kit with all the parts. If capital is limiting and labor is not, there are low cost ways of building and operating a greenhouse. If however labor and time are the limiting factor, the extra cost invested in materials and having someone else build it can usually be justified by the value of the product coming from the greenhouse.

## LIGHT AND TEMPERATURE

Managing the greenhouse to provide appropriate light and temperature conditions for quality transplants depends on understanding how plants respond to light and temperature.

### Transplant Growth Stages

Transplant or plug production is generally described in four observable stages. The first two stages account for actual seed germination and emergence. Stage one includes the starting process from seed sowing and medium wetting to the emergence of the radical or root initial. Stage two continues from root emergence until the seed/cotyledon leaves are expanded. Stage three occurs when the primary growth phase of true leaves and an extensive root system develops. Stage four is the finishing or hardening stage. In general, the recommended moisture level and temperature tend to decrease and the amount of light and fertility will increase with the progression of each stage.

## Light Management

Optimal seed germination may depend on the absence or presence of light, although most species will germinate in light or dark. Only low intensity light (50 to 100 foot candles) is necessary for the first stage of germination. High intensity (direct, full sun) is usually avoided because it causes excessive heat or rapid drying of the growing medium. Many large plug producers use germination chambers during stage one to efficiently maintain temperatures (70 to 75 F) and moisture levels. It is critical that light intensity be increased during stage two to prevent the rapid elongation that occurs, sometimes in matter of hours, as the hypocotyl (shoot) emerges from the soil. A stretched or etiolated hypocotyl will make the remainder of transplant production more difficult.

The question of whether to cover or not to cover the seeds with root medium is usually less dependent on light and more dependent on how moisture will be maintained. Covering is not essential but helps maintain moisture. Tiny seeds like petunia, begonia or trachelium are usually not covered. Additional germination medium is most commonly used when a covering is needed. The initial stages of germination are very dependent on soil temperature (as opposed to air temperature). Soil temperature is influenced by the amount of light and the rate of moisture evaporation from the root media, which is influenced by relative humidity. Evaporation from media is significantly greater than transpirational water loss by the plant in the early stages of germination and plant growth. Soil temperature is also influenced by the color and type of growing container.

When considering greenhouse light management, light intensity (brightness measured in foot candles), duration (measured in hours), and quality (color or wavelength) are all concerns. Acceptable transplants can be grown under a wide range of light intensities, but the largest (weight, not height) transplants will be grown under the highest light intensities. For late spring and summer production, shading may be necessary, not to block an excess of light, but rather to reduce heat in the greenhouse. If ventilation is not adequate to reduce temperatures, shading compounds can be sprayed on the covering or shade fabrics can be used outside the greenhouse.

The amount of plant growth will be determined by both the intensity of the light and the duration of time the light is available. For example, if light quantity were considered, the expected result would be similar plant growth if 5,000 footcandles were provided for 5 hours or 2,500 fc were provided for 10 hrs. The product of both is 25,000-footcandle hours. This helps explain why good transplants can be grown under a fluorescent shop light if the light is kept close to the plants and left on for 18-24 hrs/day. It also helps explain why greenhouse plants do not grow as much during the winter when light intensity is low and days are short. High-pressure sodium (HPS) lamps 400 watt) are most efficient for supplemental lighting in greenhouses, and the most efficient use is in early stages of growth (stage two and early stage three) when plants are small and many plants are concentrated in a small area. A well placed light will also help increase the temperature in a small area of the greenhouse.

Light quantity influences the effect of nitrogen fertilisers on plant growth. Ammonium $NH_4^+$) toxicity may occur under very low light and cold soil (less than 55F) conditions. For the plant to make use of ammonium nitrogen, there must be adequate production of sugars and carbohydrates from photosynthesis. Low light, together with low temperatures that can occur in the winter greenhouse combined with unique requirements of early seedling growth can lead to problems. A limited number of high intensity sodium lamps (HPS) in a walled off corner of the greenhouse can provide the extra light and warmth needed at a low cost.

It is generally assumed that plants get taller or stretch in lower light. For most plants, however, there is not an increase in height with lower light intensity unless there is a change in light quality, for instance due to crowding or shading by other plants or the greenhouse covering. When plants are tightly spaced, the light quality or ratio of red to far red light changes as plant leaves absorb light. This can lead to increased plant height caused by longer stem sections between the leaves. More space between plants due to larger containers or more space on the bench means shorter plants.

### Temperature Management

Relationships between day and night temperature, the 24-hour average daily temperature, and the temperature response of the plant species being grown, are important considerations to overall temperature management.

*Average Daily Temperature (ADT).* In general, the 24-hour average daily temperature will have the greatest influence on the rate of plant growth. Most plant species grow very little at temperatures less than 45 to 50 degrees Fahrenheit (F) and will continue to grow more rapidly up to a 24 hour average of 75 to 77 F. The rate of growth can be characterised as the rate of increase in weight of the plant, or the rate of formation of new leaves. Cool season vegetables like lettuce, spinach, cabbage, broccoli, etc will grow under temperatures not suitable for warm season vegetables like tomato or vine crops. The variation in growth rate with temperature also applies to cut flowers. An average temperature range for stage 3 would be 60 to 65F.

Greenhouse temperatures also have an effect on nutrient availability from organic fertilisers or compost based media. The mineralisation of nitrogen and the conversion of ammonium nitrogen to nitrate nitrogen are dependent on the activity of soil microorganisms. At soil temperatures of 55F and above, nutrient and nitrogen availability are usually adequate and increase as plant growth increases. However, the soil warms more slowly than the air after a cold spell and the recovery time for the microorganisms may be longer than for the plant. Bottom or soil heat is highly recommended for seed germination and transplants if air temperature is maintained below 55F.

*Day/Night Difference (DIF).* Considering how day and night temperatures vary can help in determining how the plant will look. When the day is very warm and the night is cool or cold, plants will be taller. If the day and night temperature are both the same, plants will be shorter than with warm days and cool nights. If the night temperature in the greenhouse is kept warmer than the day temperature by using heating at night and ventilation during the day, the plants will be even shorter. Keeping day temperatures cool (<60) rather than letting the greenhouse get very warm (>70F) will help keep transplants shorter. The relationship is referred to as DIF, or difference between day minus night.

### Relative Humidity and Condensation

Light, temperature and ventilation will also influence the moisture in the air. High relative humidity limits nutrient uptake and increases the potential for foliar and flower diseases. Condensation on the roof covering can also dramatically reduce light intensity. Ventilation to remove moisture will usually precede attempts to conserve energy or heat in the greenhouse.

### Hardening Finished Plants

The transition from the greenhouse to the field involves changes in light, temperature and wind. A gradual transition over a few days with some protection from wind and temperature but full exposure to light can increase the survival rate of transplants in the field.

## ROOT ZONE MANAGEMENT

Irrigation water quality, irrigation method, root medium, and fertiliser all interact to define the root zone. For transplant production, the usually very limited root volume and the variety of container sizes and materials also play a major factor in managing the root zone. Rapidly growing plants with limited root volumes require careful water and nutrient management to provide a healthy transplant with proportioned root and shoot growth.

### Growing Containers

Transplants can be grown in all types and sizes of containers including no container at all if soil blocks are used. Before the routine availability But what is most economical and efficient for the small grower producing a wide range of diversified crops? Efficiencies of plug production of plastics, transplants were successfully grown in wooden flats and roots were simply pulled or cut apart at transplanting. Plastic or foam trays allow for less transplant shock and ease of handling individual plants with intact root balls. But they can also add to materials cost and be a disposal problem if recycling is not an option. A decision will have to be made regarding whether germination and finishing will occur in the same container or whether germination will be done in one container followed by transplanting to a finishing container.

### Seed Germination

Germination and a longer growth period in small plugs followed by shorter finishing times in larger cells does require more attention to detail than germination in open flats for shorter times and longer times in finishing flats. However, germinating and growing in 800, 600, 500 and 400 cell trays is the most challenging and probably best left to specialty propagators. If transplants will be direct sown and finished in 200 - 128 cell or larger trays, Aplug@ production is much easier to do and might better be called transplant production.

Plug production techniques have evolved based on soilless root media and frequent applications of primarily nitrate nitrogen water soluble fertiliser. Compost based root media and less frequent application of water soluble organic fertilisers like fish emulsion can be successfully used for plug production. are partially based on specialty propagators producing large numbers of flats of fewer species and cultivars. Plug production is less an advantage with production of smaller numbers of a wide diversity of crops. These flats require very little germination root media. Crops with similar germination times should be grouped within a flat. If watering is a concern, or the time from sowing to transplant will be longer, individual 4 or 6 inch pots can be used for sowing seeds. With plugs or any germination container, the smaller the cell, the more frequent the need for water.

*Transplants*: One key distinction that can be made is whether or not the plants are stressed or treated to favor the induction of flowering. For ornamental bedding plant transplant production, the market demands miniature plants in flower even though long term garden performance would likely be favored by more vegetative transplants. Drought and low nitrogen are frequently used to control plant size and initiate early bud set. This strategy would produce a totally unacceptable transplant for cut flower production. For field grown cut flowers it is important that growth not be checked or flower initiation encouraged prematurely.

What about vegetable transplants? The answer is most likely dependent on the market and desired result. For early but limited yield that might be desired for a homeowner vegetable garden, tomato and pepper transplants in larger containers can be stressed to induce early flowering. But for field production the goal will be to put a vegetative plant out and promote rapid vegetative growth that will produce the largest yields. Production costs will likely be a primary factor influencing transplant size.

## Root Media

*Physical Properties.* While plants can be grown in all types or root media, seed germination and transplant production have perhaps the greatest requirement for a high quality, uniform medium. It is sometimes noted that coarse or large particle size media are desirable due to the need for high aeration during germination. However, most germination media are actual fine with small particle size in order to provide uniformity for filling small plug cells. Aeration is maintained instead by not over watering or saturating the root medium.

*Chemical Properties.* Nutrient levels in a germination medium need to be low but balanced. Excess soluble salts can reduce water uptake and damage young roots. Recommended electrical conductivity values are 0.4 to 0.6 mS for a 1:2 soil to water sample or less than 1.0 for a saturated medium extract (SME). The root medium pH should also be in the range of 5.5 to 6.5, although many species will germinate over a range of 5.0 to 7.5.

*Biological Properties.* The presence of plant pathogens is also a primary concern since germinating seedlings are very susceptible to root rot fungi. This is perhaps the main reason given for using soilless or peat-based media for seed germination. Compost is far from sterile but can be used for seed germination. Given these basic guidelines, the selection of root medium components and amendments together with fertiliser will likely have the greatest impact on whether transplant production is suitable for organic certification.

*Components.* Components are the materials used to provide the base media, and are present at rates of at least

10% by volume. Common components such as peat, perlite, vermiculite, coconut coir, and bark are acceptable for organic certification. Compost is the most renewable and a preferred organic substrate.

*Peat:* Peat is the leading transplant media candidate. There are many grades of peat and for potting media. Usually the best Canadian sphagnum grade available is worth the investment (Sungro, Fafard, Premier). Liming recommendations are different when more or less peat or different types of peat are added (5 to 10 lb per cubic yard range for 70% peat medium).

*Coconut Coir:* Another material like peat is coir. It is a renewable substrate like peat, but it is transported much further. It does work great for growing plants and has good water and air holding properties. The level of soluble salts (EC) would be expected to be higher than peat as would the pH.

*Compost:* Well made compost can have physical properties (air and water holding) similar to peat. However, the nutrient content would be much higher. It is also critical to recognise the difference in pH. While peat is often very acidic and requires the addition of lime, compost is often basic when first finished, but usually neutral when fully mature. We choose the mixture of peat and compost primarily based on attaining the desired pH. For the compost, we can made from garden waste, straw, hay and sheep/horse manure, a 50/50 blend of peat and compost has yielded a pH of around 6.0 which is desired for a potting media. At pH of 5.8 to 6.3, phosphorus will be much more available, particularly if there is little or no soil present. Sieving or screening the compost helps provide a more uniform product. Compost feedstocks and maturity determine pH and the pH must be tested.

When it comes to making compost for potting medium, one could refer to a process of making "designer" composts. The quantity needed is not large, and the impact on the overall farm production can be very large. Transplants play such a key role that making compost specifically for transplants makes sense.

Following is an example of a non manure compost recipe for transplant production (or compost tea). The goal is to use readily available material that would be affordable and reasonably constant across geographic regions.

1 bale straw

1 bale grass hay or grass alfalfa mix

1 bale wood shavings for bedding

1 bale (3.8 cu ft) peatmoss

6 cubic feet (wheel barrow) of soil

6 cubic feet (wheel barrow) of grass clippings (If green plants are not available use alfalfa hay.)

6 to 12 cubic feet (wheel barrow) of green plants like comfrey, weeds without seeds, green manure, etc.

We make this mixture by layering the components in a small manure spreader and then letting them come out the back in a pile. Water is added during layering and while the material is discharge from the manure spreader. This mixture reached over 140F in four days and held a temperature above 130 for one week or more. The pile was put back into the spreader for mixing. Alfalfa meal was added (25 lbs or 5 gallon bucket to about 4 yards) to assure a rapid second heat. This mixing was done in June and July when green plant material was readily available.

The mixture was kept moist and allowed to mature through the remaining part of the summer. The piles were covered and left outside for the winter. Ideally the material would be brought into the greenhouse in the fall to be ready for spring transplants. Fertilisers such as rock phosphate or greensand could be added during composting or when the compost is brought in for the winter (although it is not likely needed). These materials have low solubility and may not contribute much, but they definitely will not hurt things and end up in the field with the transplant.

*Perlite and Vermiculite:* Coarse components like perlite or vermiculite are usually added because they help maintain airspace under conditions of fine particle peat or compost and improper irrigation. However, with quality peat and compost, they may not be needed for aeration. The level of soluble salts in the compost may be high enough that perlite and or vermiculite are needed to lower soluble salt concentrations to ranges acceptable for seed germination.

*Bark and Sand:* Based on my experience to date, even aged or composted pine bark would not be recommended because of the tendency to tie up nitrogen, which usually is in short supply with organic fertilisers. Nitrogen tie up with bark is not a problem with synthetic fertilisers because it is easy to add a little more N. While sand is often still listed as a possible component to add aeration, it rarely contributes to adding pore space in already porous media. The potential for chemical and disease contaminants are other reasons to not use sand.

A likely problem with peat-based media is the low nutrient content and holding capacity. With the addition of slowly soluble or available fertilisers (listed later), and water soluble fertilisers, organic fertilisation can be accomplished. However, the addition of compost can significantly increase nutrient retention and capacity and is highly recommended.

*Amendments.* Examples of common amendments (less than 10% by volume) to peatbased media are lime, fertilisers (potassium nitrate, superphosphate, micronutrients), wetting agents or other additives. Synthetic fertilisers and wetting agents routinely added to peat-based media are not acceptable for organic certification. Wetting agents are needed primarily when peat is very dry. One option is to not allow the peat to get very dry. Another consideration is that other components like coconut coir, perlite and vermiculite wet without wetting agent. Sand also can help increase rewetting. If the peat is moistened with warm water while blending during formulation with other components, it is likely a wetting agent is not necessary. Peat does vary in wettability.

*Formulations:* The recommendation to date is for a blend of peat and compost or peat-based soilless medium and compost. The current greenhouse standard unit of measure for amendments is pounds added per cubic yard of medium (3′ x 3′ x 3′ = 27 cubic feet). Amendments are not essential but can include: blood meal (nitrogen), rockphosphate (phosphorus and calcium), greensand (potassium and micronutrients) and other sources in the table provided. Elliot Coleman recommends combining equal parts of these three (although he does not state equal parts by volume or weight, we assumed volume), and then adding a rate equivalent to 14 lbs per cubic yard to a peat or peat-based medium and compost (50/50 by volume) blend. He recommends letting the blend sit for a month or more before using, although he does not state that this is essential. This or similar formulations were very effective for us with edible flowers and herbs.

## Water Soluble Fertilisers

With the use of synthetic water-soluble fertilisers and mechanical injection devices (proportioners), fertilisation is one of the easiest aspects of large scale greenhouse production. For Organic certification, there are materials available but few specific, published recommendations. With a well prepared root medium, only nitrogen (N) will likely require management during production. While some phosphorus (P) and potassium (K) can be added, it is N that has the greatest effect on plant size. (Assumes calcium (Ca), magnesium (Mg) and sulfur (S) are in the root medium or water.

*Fish Emulsions.* We have used soilless peatbased medium and coconut coir without synthetic fertiliser added and only fish emulsion (5-1-1) used as fertiliser for periods of several months up to almost a year without any apparent nutrient deficiencies. Use of fish emulsion as a sole fertiliser for potted plant production has been well demonstrated by others, many years before we tried.

For a 5-1-1 liquid formulation, a dilution rate of 1 to 100 (1 gallon in 100 gallons or 6.4 ounces in 5 gallons) will provide 500 ppm. (1% liquid equals 10,000 ppm) A dilution rate of 1to 1000 will provide 50 ppm N. The best rate is in this range and will be determined by the N levels in the root medium, the desired growth and the frequency of application. When a fertiliser injector is available and rapid growth is desired, fertilisation at every irrigation with low rates like 50 to 100 ppm N is common. However, with products like fish emulsion that have a fragrance to contend with, fertilisation at biweekly or monthly frequency is preferred. Interval fertilisation also prevents over fertilisation. Fertilisation can be delayed until the plants show signs of lower leaf yellowing or slowed growth that indicates more N is needed. A final shot of N just before transplanting may be beneficial.

### Irrigation

*Water Quality.* Water quality refers to the concentration and species of chemical elements found in water that make it suitable or unsuitable for irrigation. In general, the shorter the crop time, as for transplants, the less effect water quality will have on plant growth. However, it is also true that the smaller the growing container and the more frequent the irrigation, the greater the effect of water quality. It is important to have your water analysed and to know what desired or undesired elements are present and how they will effect your choice of root media and fertilisation.

*Method of Application.* The success or failure of plants in a greenhouse is most often attributed to the person handling irrigation. The primary questions to be addressed are When to water, How much or what volume to water, and What method of application to use.

When to water will change over time. In early stages of seed germination it is critical that moisture availability be very high. Once root growth begins, moisture levels are gradually reduced over time. In finishing stages transplants may be wilted on purpose.

How much to water refers to whether or not to partially moisten the root media, thoroughly moisten the root medium, or water enough so excess drains from the root medium (leaching). Standard rules of watering to thoroughly saturate or leach at every irrigation do not apply to seed germination or transplant production. Root medium air space in shallow containers is maintained more by how much water is added than by the particle size of the root medium. Leaching is also not recommended unless soluble salt levels are excessive. Usually the intent is to retain as much of the nutrients in the root medium as possible, particularly with organic production methods and the small root medium volume.

Choices of how to water include overhead watering using hose and breaker or sprinklers verses subirrigation by flooding from below. Making sure the foliage is dry at the end of the day is important in the later stages for disease control. The best system may be one that combines overhead irrigation in the early stages and subirrigation in the later stages.

If water is going to be used to regulate plant size, it is important to recognise that there is more than one way to grow plants dry. One way is to water plants thoroughly and then allow them to dry until wilting occurs. This method does not reduce plant height nearly as much as frequent, light applications that do not saturate the root medium and keep the plant under a constant low level of drought stress.

### PESTS AND DISEASES

It is totally within the realm of possibility that if your farm or greenhouse is relatively isolated, the greenhouse has been empty and at low temperatures over the winter, you have not over wintered plants in the greenhouse, you start strictly with seeds, as opposed to purchased cuttings, your In the first year of our research with edible flowers and herbs, we battled the white fly and thrips primarily by releasing purchased predators or parasites. The second year we used what we learned previously to keep the whitefly and thrip within acceptable levels, but the aphids were a major problem. We used more predators and occasional sprays, which were suitable for organic certification but were as hard on the purchased predators as they were on the pests. The third year we did not have any major outbreaks but we had some problems with spider mites and potentially devastating problems with mealy bug. For the potting medium doesn't harbor any noxious pests and diseases, you don't over water or over fertilise your transplants, and you minimise crop time in the greenhouse, that you could never see or experience any of the following pests and diseases. The best insect and disease management program in the greenhouse is based on a strong defense and a well thought out plan.

A good starting recommendation is to use the best cultural methods as outlined in the previous two sections and grow a healthy transplant. The much published and referred to pest and disease control triangle includes the environment, the host and the infesting or infecting agent. Start with a growing environment providing proper light and temperature conditions that favor growth of the plant and minimise the growth of the insect or disease. For example, warmer temperatures mean faster growing transplants, but will also mean greater potential for pest populations to develop. The third part of the triangle is knowing as much as possible about the pest or disease.

### Pests

*Identification.* Based on experience with large, commercial greenhouse operations the ten most common greenhouse pests could be rated this way. The three that present the largest, routine problems are white fly, thrip, and fungus gnats/shore flies. The next three that are occasional problems or more crop specific problems are aphid, spider mite, and leaf miners. Less common but still a problems with certain crops are mealy bug, scale, chewing caterpillars and slugs. previous mentioned research, we were located in a large greenhouse range of over one acre with many different research projects and crops. In general all the pests were present and chemical applications were heavy.

*Management.* Perhaps one of the best recommendations is to minimise transplant crop time. Schedule carefully so plants are not in the greenhouse any longer than necessary. The longer the production and or holding time, the greater the probability for uninvited guests to move in.

An option in the greenhouse, particularly for short term crops like transplants, is to use either protectant or biorational pest control sprays: (sprays based on naturally occurring control agents or competitors). Examples of sprays acceptable for organic certification by at least some certifying agencies include insecticidal soap (Ex. M-Pede), Buavaria fungus (Ex. Botanigard), neem based products (Ex. Neemx, Azatin), garlic or hot pepper barriers, compost teas and herbal or mineral preparations.

### Diseases

While healthy growing transplants are often resistant to

infection by the ever present root rot fungi, the initial phases of seed germination and the early stages of seedling growth are perhaps the most susceptible parts of the plant life cycle. If root rot pathogens are present at high levels, plant losses can be complete.

Root rot pathogens potentially present in root media in conventional greenhouses are generally controlled by careful selection of the root medium, watering practices that minimise over or under watering, and use of selective fungicides that target a fairly narrow range of fungi. The use of selective fungicides is very important in routine large scale greenhouse production and widely practiced. This practice is not acceptable for organic transplant production. The techniques of steam or chemical sterilisation (kill everything) or even pasteurization (kill most of the bad guys) are no longer commonly practiced in most commercial greenhouses. While there are authors that recommend baking or heat treatment of seed germination media, the practice is not practical to accomplish and of questionable value.

Germination and transplant media formulated from sphagnum peat, perlite and vermiculite will generally not require any additional treatment. The level of innoculum present in peat is typically low and perlite and vermiculite are sterile as a result of the manufacturing process. If compost is to be added, the composting procedures should be tested for the suitability to clean the compost of seed (root rot) pathogens. When the heat of composting (130-150 F) is maintained for adequate time (2 to 3 days) and the entire pile is heated at some time via mixing and reheating, the resulting compost should be fine for germination media assuming the quality and maturity are satisfactory. Identification.

The most common recognised greenhouse root rot pathogens are Pythium and Rhizoctonia. Also problems at times with specific crops are Phytophthora, Fusarium and

Theleviopsis. The most common recognised greenhouse foliar fungal pathogens are powdery mildew and Botrytis. The most common bacterial pathogens in the greenhouse are Erwinia, Pseudomanas and Xanthomonas. The most common viral pathogen in commercial greenhouse currently is impatiens necrotic spot virus (INSV) which is transmitted by thrips. When soil is used in the greenhouse, nematodes could also be a concern (not likely).

*Management.* The best control for root rots is proper formulation of the root media, proper watering, and fertilisation practices that favor healthy, undamaged roots. Foliar fungal pathogens are best minimised by keeping relative humidity in the greenhouse low (<80%) and water off the foliage. Reducing relative humidity may require venting warm moist air out of the greenhouse late in the day. For bacterial and viral infections, rouging potentially infected plants as soon as possible, minimise splashing water that can spread bacteria, and controlling aphids and thrips that can spread viruses are recommended. The use of insecticidal soap as a control agent for powdery mildew has been suggested.

**REFERENCES**

Byczynski, Lynn, "Growing great transplants", *Growing for Market*, February 1993.

Paul V. Nelson, *Greenhouse Operation and Management* (5th edition), Prentice Hall, Inc., Upper Saddle River, 1998..

Parnes, R., "Fertile Soil", *A growers guide to organic and inorganic fertilizers*, AgAcces, Davis, CA, 1990.

Robert Aldrich and John Bartok, *Greenhouse Engineering*, Northeast Regional Agricultural Engineering Service, 1989.

# 7

## Genetically Modified Plants and Organic Agriculture

Genetically modified pest protected plants or Genetically Modified Plant Pesticides (GMPPs) have radically changed agriculture in the few years they have been commercially released. The effects that this and other techniques of genetic engineering in agriculture will have on farming are not yet apparent. The organic community is concerned about the impact that this technology will have both on their immediate growing conditions and the long term sustainability of agriculture.

Genetically engineered organisms in general and GMPPs in particular are considered incompatible with organic production for a number of reasons. Among these are:

(1) The recombination of DNA changes the molecular structure of the target organism by biochemical—as opposed to biological—means.

(2) Genetic engineering develops and produces products in a way not possible or practical with traditional techniques used in organic production and processing.

(3) Gene transfers between organisms not capable of exchanging genetic materials by sexual reproduction are intentionally produced to develop a desired trait.

The process of splitting DNA, and deleting, inserting or changing the position of genes involves the use of various compounds. By these means, rDNA techniques take advantage of the means by which organisms repair damaged genetic material. The molecular structure of rDNA is different from the structure of both the donor and recipient DNA in ways that would not be possible through a natural biological process.

Genetic engineering's promoters often state that rDNA techniques differ only slightly from the classical plant breeding of pest or disease resistant plants. However GMPP corn differs in a number of ways from a hybrid variety. Traditional breeding makes use of natural reproductive mechanisms to select desirable traits expressed by a given set of genes from among the many traits present in a population of closely related organisms. While human manipulation is part of the system, the use of natural reproductive mechanisms limits the use of traditional breeding to a single species or to closely related species that can crossbreed.

Genetic engineering is clearly distinct from those advances based on the extension of traditional breeding, propagation, and improvement programs. Hybridisation, tissue culture, in-vitro fertilisation, and transconjugation are all applications of biotechnology that stop short of genetic engineering. While these applications can create problems for genetic diversity and the agroecosystem, such techniques have long been a part of organic farming systems and are well understood.

By contrast, genetic engineering breaches the natural reproductive barriers between or among species. The laboratory procedures involved in rDNA methods isolate genes from one organism and then transfer those genes directly into another organism. This makes it possible to transfer genes from any organism on the planet to virtually any other organism, regardless of their sexual compatibility or the evolutionary distance between them.

Genetic engineering can also be used to transfer genes between members of the same species, in order to duplicate or delete that gene or alter its information content. All organisms modified using recombinant DNA techniques are called "genetically engineered" or "transgenic," regardless of whether recombinant manipulations involve the transfer of genes between highly divergent species or between members of the same species.

## GMPPS AND SUSTAINABLE AGRICULTURE

There are several different cultivated plant species. This has enabled those plants to express the same toxins produced by Bt. The organic community has heard a number of arguments both supporting and opposing genetic engineering's compatibility with organic principles. Among the arguments made in favor of allowing GMPPs in organic production include:

- GMPPs are simply an extension of traditional breeding techniques for developing resistant varieties.
- Farmers can reduce pesticide use by planting varieties engineered for pest resistance.
- GMPPs are based on natural toxins already used in organic agriculture.
- GMPPs can provide organic farmers and processors with more tools to increase the productivity and lower the costs of organic production and processing.
- Organic producers, processors, and handlers will soon lose commonly available inputs as GMPPs displace the original sources of these natural toxins.

These assertions are not well supported by any empirical evidence, and there are a number of reasons that GMPPs appear incompatible with organic principles, such as:

— GMPPs are too new to be able to predict the long-term effects on human health and the environment.

— The amplification of the introduced toxin (high dose) and synergy with the natural toxins and allergens occuring in plants may compromise food safety in ways that cannot be predicted based on the toxicological profile of the individual constituents.

— GMPPs threaten the biodiversity of both the host plant and donor organism.

— Certain products of genetic engineering—combined with a liberal interpretation of the organic standards—would increase, rather than decrease, chemical usage on organic farms. Of particular concern are herbicide tolerant crops.

— GMPP field tests have not gathered adequate data to conduct a valid risk assessment. In particular, protocols reviewed do not evaluate the potential impacts of GMPP crop residue on organisms such as beneficial insects and microorganisms that play a vital role in humus formation and cycling crop plant nutrients.

— GMPPs have the potential to select resistant target pests by the widespread release of naturally occurring toxins in more persistent and toxic forms.

— GMPPs could create new weeds or pests not easily controlled through the horizontal transfer of genetic material.

GMPPs are considered 'synthetic' for the purposes of setting an organic standard. The transfer of genetic material under natural conditions is usually limited to within a species. Transfer between species takes place within nature, and is means by which many cultivated varieties of plants have developed. Transfers between more distant relatives may take place through a variety of means. These transfers are known as gene flow. The frequency and significance of the flow of genetic material under natural conditions is the subject of academic debate. These uncontrolled transfers are often unstable or result in mutations that are not reproductively viable. GMOs are, by contrast, not random and must be robust enough to bring the desired traits to fruition.

Other researchers dismiss as invalid or—at best—deficient, any evidence that horizontal gene transfers occur under any natural conditions as invalid, or deficient at best, thus supporting the case that such the transfer of genes between species is 'unnatural,' and therefore synthetic. If horizontal gene flow is common, this increases the probability—and therefore the risk—of ecological damage that results from GMPPs crossing with wild relatives to create new weeds. On the other hand, if gene flow is uncommon, then transgenic organisms are more clearly unnatural, or 'synthetic.' It is possible for GMPPs to transfer genes, and for those transfers to have ecological repercussions. Given that scientists disagree over the probabilities and consequences of such an occurrence, the probability of horizontal gene transfer and the consequence of such an event should not be the main consideration to determine whether or not GMPPs can co-exist with organic farming systems.

### Conservation of Bt

*Bacillus thuringiensis* is often called the ideal pesticide. It is narrow-spectrum, with each strain effective against a family. One reason that farmers give for not using Bt more often is its selectivity. However, its selective nature makes it an attractive option for farmers who rely heavily on beneficial organisms as part of their pest management strategy. Bt is a naturally occurring soil organism that is found throughout the world, thus not likely to upset soil microbiology. The toxins produced are essentially non-toxic to humans and other mammals. It is effective because it is highly toxic in low doses to the target species.

Bt is most effective against pests that are both highly susceptible and have feeding habits that cause the foliar applied spores required to be ingested. To be effective, the Bt must both be ingested, and the target pest must metabolise the protein to activate the toxin. An insect that does not metabolise Bt will be immune to it as a pesticide, or resistant. An insect may be susceptible to one strain of Bt and resistant to another. This is usually related to the specific toxins produced by the strain of Bt, known as the d-endotoxin. Four different groups of crystal proteins have been identified based on their amino acid structure. CryI and CryII are active against lepidoptera. CryI crystals are found in commercially available strains aizawai, berliner and kurstaki. The kurstaki strain also contains CryII structures. Over 100 species of lepidopteran pests have been identified as susceptible to the kurstaki strain. Strains active against coleoptera, such as tenebrionis, contain CryIII protein structures. The commercial strain used to control mosquitoes—israelensis—contains the CryIV class of crystals.

The focus of most Bt research conducted over the past ten years has been on recombining genes that express the various Cry toxins into organisms of different genuses, often across different kingdoms. These new techniques have had a number of implications for the toxicity, persistance, and even mode of action of the Bt toxin. Transgenic plants have been engineered to express the toxin directly, rather than depend upon the Bt organism to be applied to the foliage to deliver the toxins. While it is possible to introduce multiple genes into another organism, transgenic products are usually engineered to express a single toxin.

GMPP policy based solely on experience with Bt as the donor organism will not produce valid results for other toxins expressed by other donor organisms. Transgenic Bt is a prototype for other GMPPs. Numerous other natural toxins are being screened as potential insecticides. The same technique could be extended to a number of other toxins that have greater environmental or human health impacts than the Bt toxin. These toxins could be derived from plants such as pyrethrum, derris and qubé (rotenone), and Nux vomica (strychnine); from viruses, bacteria or other microorganisms, or arachnid-derived neurotoxins. Any model used to evaluate GMPPs should take into account the specific host-pesttoxin interaction and the ecological system that is exposed to the release of that organism to provide a valid assessment of the impacts.

Transgenic corn, cotton, and potatoes were introduced commercially in 1996. The release of these products has raised a number of issues from the unknown effect of concentrated toxins, increased persistence, and the potential for the Bt gene to escape to wild relatives of the receiving organisms. The very lack of practical experience, knowledge, and predictability of the outcomes of introducing GMPPs into the ecological system is itself a factor in questioning the technology's sustainability. Since the outcomes of these manipulations cannot be predicted as fully as the outcomes of traditional breeding, there is tremendous uncertainty associated with genetically engineered crops both in terms of environmental impacts and food safety.

Risk assessment requires data on potential outcomes, their probabilities, and the costs associated with those outcomes. Since the outcomes from adopting the technology can't be predicted ex ante, and the probabilities and costs associated with those outcomes are indeterminate, the risks cannot be determined and risk-benefit analysis is not a valid model to make scientific decisions. Add to this the likelihood that some of the potential outcomes are irreversible, another model for technological decision-making is more appropriate.

Organic agriculture has been more conservative in adopting most new technologies than other farming sectors. However, in some cases, organic farmers have been innovators, pioneers and early adopters of biopesticides. Organic farming standards are based on the premise that synthetic materials are prohibited, unless they are explicitly allowed on the basis of the following criteria:

(1) the potential of such substances for detrimental chemical interactions with other materials used in organic farming systems;
(2) the toxicity and mode of action of the substance and of its breakdown products or any contaminants, and their persistence and areas of concentration in the environment;
(3) the probability of environmental contamination during manufacture, use, misuse or disposal of such substance;
(4) the effect of the substance on human health;
(5) the effects of the substance on biological and chemical interactions in the agroecosystem, including the physiological effects of the substance on soil organisms (including the salt index and solubility of the soil), crops and livestock;
(6) the alternatives to using the substance in terms of practices or other available materials; and
(7) its compatibility with a system of sustainable agriculture.

Given that GMPPs are synthetic, then these criteria can be applied to evaluate their compatibility with organic standards.

### Detrimental Interactions

The ecological consequences of releasing GMPPs into the environment are not obvious, and numerous effects of genetically engineered products cannot be predicted with accuracy prior to their commercial release. Bt toxins expressed by other host organisms may have different interactions with the environment than the non-GMO bacteria. The interaction of different strains can have a synergistic effect. These strains vary widely in their host range. Given the complexity of the biochemistry of different plants, the substances naturally exuded by plants may cause detrimental interactions with the Bt toxins.

**Toxicity and Mode of Action**

There are a number of ways in which genetic engineering can increase the toxicity and enhance the modes of action for many familiar and novel agricultural products. Transgenic plants engineered for resistance to insects and diseases do so by the production of toxins. Many transgenic plants are commodity crops grown as animal feed. Genetically engineered feed may also present health risks to animals through increased levels of toxins, altered nutrient composition, and nutrient availability. For other products, the toxicity studies have not been performed. GMPPs and their products may result in allergens that cannot be predicted.

**Environmental Contamination**

One measure of the adverse effects of genetic engineering on the global environment is genetic pollution. Hybridisation between genetically modified crops and wild relatives threatens the loss of biodiversity and the natural germplasm on which breeders depend for new traits. Recombination also increases the risk of the generation of new viral strains.

When a GMPP is introduced into the environment, it may upset the farm's natural balance. New organisms released into the environment may mutate or interbreed with weedy species, and become pests, pathogens or weeds themselves. Many applications of biotechnology—hybridisation and tissue culture, as well as genetic engineering—threaten the diversity that maintains the stability of agricultural systems.

Genetically engineered organisms may displace plants or animals from their niches, thus making them endangered species or even driving them to extinction. The escape of a weedy GMPP is a function of the following factors: (a) the ability of the organism to survive; (b) the organism's propensity to reproduce; (c) the introduced species interaction with other organisms; and (d) the nature of that interaction.

### Human Health Concerns

GMPPs may directly create human health and environmental problems. Any genetically engineered product has a dimension of risk that cannot be predicted before its use. Because genes can affect more than one trait (pleiotropism), it is impossible to predict ex ante the effects of genetic manipulation. Models developed for the evaluation of risks of synthetic chemicals do not necessarily provide a meaningful evaluation of the risk of genetic engineering. For example, insect and disease resistant crops may produce more potent natural toxicants at greater levels than those currently found in plants. New models need to be developed and tested to assess toxicity, allergenicity and antigenicity of expressed products. Also, the stability and form of the Bt toxin may change over several generations that cannot be predicted without long-term research.

### Effects on the Agroecosystem

Methodologies to evaluate the environmental impact of the release of live genetically engineered organisms are in their infancy. The effects of such organisms on the ecosystem have been studied only a short time. Release of GMPPs into the soil environment could result in "(i) displacement of existing species, (ii) major changes in community structure and function; (iii) perturbation of ecological balance; (iv) accumulation of toxic metabolites; or (v) increased microbial activity due to nutrient input".

The most serious concern is the risk that such products could accelerate the failure of Bt through resistance. A more toxic and persistent form of Bt that coincides with the release of transgenic plants theoretically could increase selection pressure for Bt resistant strains of insects. Laboratory resistance to Bt was first reported in the Indian meal moth (Plodia interpuntella). The diamondback moth was the first species to show resistance in field conditions.

Resistance is not inherent to transgenic Bt. Prolonged or constant exposure of the target population to the toxins possibly provides for much greater pressure to select for resistance than with standard foliar Bt. Continuous exposure to the Bt toxin created the selection pressure for resistant variants of pests. Extensive and intensive exposure of pests to Bt through transgenic plants may cause widespread pest resistance.

Crops have co-evolved with insects to exude various compounds to deter, repel, or inhibit feeding by pests. Insects overcome these natural defenses in host plants by selection pressure that favors those insects that are able to detoxify those compounds. The ability of these insects to detoxify compounds that plants express increases their ability to damage crops or virulence. This same genetic mechanism also serves as a vehicle for those insect populations to evolve with insecticide resistance as a common trait. Data obtained from crops engineered to express the Bt toxin indicated that resistance can emerge in as few as seven generations. This may take as little as two years in some climates.

The maintenance of non-transgenic refugia has been proposed as one resistance strategy. This involves farmers not planting GMPPs for a portion of the acreage in a given crop. It is not possible to predict how large a refuge area is needed before the organism is released. The only way to determine the optimal size of refugia is under actual field conditions. By not cultivating GMPPs, organic farmers increase the acreage available as refugia for susceptible gene pool for target pests.

Organic farmers may lose Bt as an effective crop protection tool as a result of accelerated resistance from the release of transgenic Bt. Experience with a series of chemicals suggests that if one chemical is used as the primary or only means to controlling a target insect, selection pressure for resistance to that pesticide will be increased. This is consistent with modeled systems that showed GMPP plants could accelerate Bt resistance in target pests.

As molecular ecology and evolutionary population dynamics advance, the risk assessments conducted to evaluate the release of these organisms are not based on sound scientific principles. Most risk assessments are based on small, confined, protected populations that are isolated from both competitors and wild relatives. These experiments are not valid models of the actual field conditions in which crops are grown. It is not scientific to call GMOs 'safe' because 'nothing happened' under these conditions. Addressing the impact of genetically engineered Bt on the agroecosystem will require large-scale experiments that examine aspects of both molecular biology and adult insect behavior that have not been of concern to researchers in the past.

**Alternatives**

None-engineered Bt is obviously a viable alternative, one that may be lost as a direct result of the introduction of GMPPs. Bt can be conserved through well-timed applications based on scouting. In addition, there are a number of biological and cultural methods for every one of the target pests of GMPPs. Numerous cultural and biological practices offer options besides GMPPs, Bt, and synthetic pesticides. Among these are rotation of non-host crops, intercropping of non-host crops, habitat management to maintain beneficial organism populations, augmentive releases of beneficial organisms, classical breeding of pest resistant varieties, crop nutrient and water management, timing of planting and harvest, and mating disruption.

Many non-recombinant methods show great promise in improving the efficacy of Bt. Discovery and selection of new strains for novel proteins and field stability through heat and UV resistance show promising developments. Research on adjuvants is one area that has shown great progress, albeit one with little published data. Bt manufacturers have successfully applied transconjugation techniques resulting in a strain with greater efficacy and a broader number of target species. Sexual reproduction of two Bt strains can produce offspring with a complementary set of genes to express the toxins produced by both parent strains.

### Compatibility with Sustainable Agriculture

There are a number of reasons to be skeptical of any claims that the technology is compatible with any system of sustainable agriculture. Time is the only test that will determine the long-term sustainability of GMPPs. The NAS would do well to address these long-term questions about the ecological impact of the technology. To do so requires scientists to develop a methodology based on a whole systems approach to agriculture. Proper evaluation of the technology needs to recognise the uncertainty of outcomes and irreversibility of the consequences.

— Researchers need to design experiments to assess the long-term effects of the use of GMPPs and their products on human health and the agroecosystem.

— Data to review the effects of such products on human health and the environment needs to be over a time of several growing seasons.
— Models to evaluate the data need to take into account the uncertainty and irreversibility in a way that risk-benefit analysis fails to do.

## RISK MANAGEMENT PLANNING FOR CONTAMINATION

In many ways, primary production is an exercise in risk management. One objective in managing a production system is to minimise the negative impacts that markets, finances, weather conditions, pests, diseases and other factors may have on the viability of that system. A risk that certified organic producers need to manage is that of contamination of their land or crops with prohibited chemicals or genetically-modified organisms (GMO's). This Agnote provides a brief overview of risk management planning for contamination issues and an example plan that can be modified to suit a producer's specific situation.

### Contamination Risks

In some situations, the risk of contamination from activities on adjacent land may be minimal. In others, such as intensive horticultural production or areas where aerial spraying occurs, there can be a significant risk of contamination from spray drift. Treatments applied by government authorities to control noxious weeds and pests can also pose a risk to organic properties. For many producers, there may be a risk of contamination associated with the import of compost, mulch or other materials onto the organic property.

Historically, the risk to organic properties has related to contamination with pesticides not permitted under organic standards. The possibility of contamination with GMO's poses a new potential risk to organic producers as GMO crops are introduced to agriculture.

### Consequences of Contamination

Contamination of a certified organic property with substances or materials that are not permitted under the certification standards, can result in suspension of the certification and withdrawal of any affected produce from the marketplace. This has implications for the integrity and viability of organic enterprises, many of which depend on their organic certification for improved market access. Most organic producers are therefore likely to benefit from some analysis of the contamination risks facing their enterprise and the development of at least a basic plan to minimise and manage those risks.

### Risk Management Planning

Risk management planning can be applied to any type of risk and any type of enterprise. It involves a series of steps through which risks to the enterprise are identified, analysed and treated, then monitored to determine the effectiveness of the treatment. Developing a contamination risk management plan should not be difficult for most producers, and will provide some peace of mind that potential risks to the organic enterprise have been addressed appropriately.

The steps to risk management planning are outlined briefly below. For a more detailed description of how to work through each step, using real-life examples, visit the web site listed under 'Useful information sources'.

### Planning Steps

*Identify all the risks*

— what are the contamination risks, in terms of the types of contamination and the likely sources ?

*Analyse each risk*

— what is the likelihood of the risk occurring ?

— what would be the consequences if it occurred ?

The likelihood and consequence of a risk combine to give an overall risk ranking. Suggested rankings for use in typical risk management planning are shown in Table 1. In reality, contamination of an organic property or product with prohibited substances will usually have major consequences, e.g. withdrawal of the property and product from organic certification. Because of this, the ranking of individual risks is not really useful. All contamination risks are automatically ranked highly and so need to be addressed.

*Table 1. Typical risk rankings.*

| | Consequences | | | | |
|---|---|---|---|---|---|
| *Likelihood* | *Insignificant* | *Minor* | *Moderate* | *Major* | *Catastrophic* |
| Almost certain | High | High | Extreme | Extreme | Extreme |
| Likely | Moderate | High | High | Extreme | Extreme |
| Moderate | Low | Moderate | High | Extreme | Extreme |
| Unlikely | Low | Low | Moderate | High | Extreme |
| Rare | Low | Low | Moderate | High | High |

If a certain risk has no likelihood of occurring in a particular situation, it can still be included in the plan to show that it has at least been considered. Once the contamination risks have been listed, they are evaluated as to which ones require special treatment, and which ones can be controlled easily through day-to-day management. A spray drift risk would warrant careful planning and control to avoid any potential impact on the organic farming enterprise.

*Treat the risks*

Risks can be treated in one of the following ways:

— *Avoid the risk:* e.g. avoid unwanted residues of conventional fertilisers by using fertiliser spreaders that are dedicated to organic production.
— *Control the risk:* e.g. supervise harvest to ensure produce from noncertified land is not mixed with certified produce.
— *Transfer the risk to someone else:* e.g. obtain insurance against contamination if possible or use only certified organic inputs (transfer the risk to the insurer or input manufacturer respectively).
— Retain the risk if it cannot be avoided, controlled or transferred cost-effectively—e.g. keep purchasing compost ingredients from non-organic sources, but be prepared to dispose of any (and suffer the cost) if they are found to be unacceptably contaminated.

### Communication

Communication and good relations with neighbours, councils and authorities are valuable components of a contamination risk management plan. Those people need to be informed about the basic requirements of organic production and the possible consequences of their actions.

### REFERENCES

Rissler, J. and Mellon, M., *Perils Amidst the Promise: Ecological Risks of Transgenic Crops in a Global Market*, Cambridge, MA: Union of Concerned Scientists, 1993.

Roush, R.T., "Can We Slow Adaptation by Pests to Insect Transgenic Crops?", in G.J. Persley (ed.) *Biotechnology and Integrated Pest Management*, Oxon, UK: CAB International, 1996.

Tabashnik, B.E., N. Finson and M.W. Johnson, "Managing Resistance to *Bacillus thuringiensis:* Lessons from the Diamondback Moth (Lepidoptera: Plutellidae)", *Journal Economic Entomology*, 1991.

Wolf, K., "Gene Transfer Between Organelles and the Nucleus in Lower Eukaryotes", in E.R. Schmidt and T. Hankeln (eds.) *Transgenic Organisms and Biosafety*, Berlin and New York: Springer, 1996.

# 8

# Weed Control in Organic Farming

Weeds are part of our farming systems and can have beneficial uses. If we are in the business of profitable organic crop and pasture production, then we need to manage weed species so that desirable plant species have the advantage. This requires an integrated weed management approach.

## EFFECTS OF WEEDS IN AGRICULTURAL SYSTEMS

### Positive Effects

— Some deep rooted weed species can bring nutrients to surface soil that would otherwise not be available to crop or pasture plants,

— Weeds can be indicators of soil conditions,

— Weeds can provide habitat for some beneficial insect predators,

— Weeds can contribute to biodiversity; it may be a more stable productive system if there are low population levels of a diversity of species, rather than large numbers of one or two plant species.

### Negative Effects

— Competition with crop and pasture plants and the use of available moisture, nutrients and light,

— Weeds can be host plants for pests and disease,
— Weeds can contaminate the end product of a system, eg. wool or grain.

## SOIL FERTILITY AND WEED POPULATIONS

There is the relationship that can exist between the level of soil fertility and the types of weeds that exist at a specific site. There are two common views that can help to explain this relationship.

The first is that with increasing soil fertility, weed problems can be reduced through the increased competitiveness of crop or pasture species. Whilst this may occur in many situations there will be occasions where weeds can readily adapt to levels of high soil fertility. As a general rule enhancing soil fertility when combined with best management practice for crop and pasture production, provides the productive plants with the competitive advantage.

The second view is that specific weeds can indicate specific soil problems. For example, research into soil acidification has revealed that there are weed species that can become predominant in some areas with lowering soil pH, such as sorrel (*Rumex acetosella L.*). Ameliorating soil acidity with lime can improve overall soil fertility, and when combined with productive pasture species, act to reduce the population of less desirable species.

### Integrated Weed Management

Integrated weed management involves the bringing together of a range of practices to manage a specific problem plant. There are four key principles that are fundamental to the integrated approach in organic systems.

i. Knowledge of soil, crop and pasture plant systems,
ii. Knowledge of weed species and how they may affect soil, crop and pasture systems,

iii. Knowledge and appropriate use of available management options for weeds,

iv. Monitoring the impact of management on the weed species and on the soil, crop or pasture.

## DEVELOPING A WEED MANAGEMENT PLAN

### Know the Farming System

A successful weed management plan involves accurate knowledge of the crop or pasture system that is being operated. For example, a grazing enterprise requires high quality and vigorous pasture species (both legumes and grasses) to provide feed for livestock on a year round basis. Weed management in permanent pasture focuses on the maintenance of healthy soils and productive species that can compete effectively with less desirable plants.

A cropping enterprise requires prepared paddocks so that the crop plants have the advantage over other plants and can yield to their optimum level. The principles of healthy soils and vigorous species still apply in cropping enterprises but there are additional opportunities to use a wider range of management methods to reduce the influence of weeds.

### Know the Weed

It is essential that the characteristics of the weed species are known if a plan to manage the species is to be successful. Preliminary classification of weeds includes whether they are annual, biennial or perennial, and how and when they reproduce. Paterson's Curse *(Echium plantagineum L.)* is an annual weed, germinating and setting seed within one year. Prickly lettuce (*Lactuca serriola L.*) is an example of a biennial weed, germinating and establishing in the first year, and setting seed in the second year. Sorrel is an example of a perennial weed that can reproduce by seed production but also by vegetative reproduction from shoots on lateral roots.

It is important to know how weeds affect the crop or pasture enterprise. For example, Paterson's Curse can be a problem plant within a pasture system, competing effectively with other more productive species but is generally not a problem weed in a cropping system because it can be managed through timely cultivation and appropriate agronomy. Similarly, annual ryegrass is generally not considered a weed in a pasture but it competes vigorously with crop plants.

### Management Options

There is a wide range of management options available to producers to manage weeds in organic crop and pasture systems.

a) *Grazing animals:* In pasture systems the most important tool will be the strategic use of the grazing animal to manage problem species. Knowledge of the grazing behaviour and plant selection ability of animals is important when planning a weed management program that utilises the grazing animal. For example, goats can be a useful addition to a mixed farming enterprise because they find some plant species highly palatable irrespective of the plant's stage of growth, eg. blackberry (*Rubus spp.*) and prickly lettuce.

b) *Cutting hay or silage:* This practice can successfully prevent seed set of less desirable species by removing the seed-head before maturity. When this method is used in conjunction with appropriate grazing practice and soil fertility programs, the more productive pasture species are given the advantage.

c) *Observation and manual removal*: Regular observation and manual removal of potential problem weeds in permanent pastures before they establish is an essential addition to the toolkit. A few spear thistles (*Cirsium vulgare [Savi] Ten.*) in one year may require a relatively small time investment to manage by hand. If let to set seed the population can expand to hundreds of plants in the following year, presenting a far more significant problem.

d) *Cultivation:* Cultivation is still an important method to remove weeds during the cropping phase but the fragile nature of most Australian soils means that cultivation should be used sparingly in any crop program. Cultivation is only one tool amongst many, and shouldn't be considered in isolation from a comprehensive integrated weed management program.

One example where light cultivation can be very effective is the practice known as 'autumn tickle'. This involves a light cultivation after early break rains to stimulate ryegrass germination so that weed numbers can be reduced before crop establishment.

e) *Other methods*: Other methods to manage weeds in mixed farming systems include:

— Rotation between different crop species,

— Incorporation of a legume break/forage crop between the pasture and crop phases,

— Selection of clean seed, vigorous crop and pasture cultivars and higher seeding rates,

— Harvest technology that captures weed seeds from the header, reducing the potential for increased weed populations,

— Blind harrowing (cultivation after sowing but before emergence of the crop) can be used but this method can in some instances act to stimulate weed germination,

— Post-emergent cultivation between crop rows is sometimes practised but requires specialised equipment,

— Biological control agents.

In cropping systems there are additional opportunities to manage weeds prior to crop establishment. Since grass weeds compete with most crop species, management of these weeds during the pasture phase is essential. Cutting hay or silage, mowing, or strategic heavy grazing to reduce seed set can all significantly reduce the seed-bank of these species. Management to reduce grass weed seed-banks should ideally occur at least two years prior to crop establishment.

### Monitoring the Impact of Management

The final component of an integrated weed management program is monitoring the impact of different methods on the crop or pasture system. Having put some weed management measures in place, for example, cutting silage or mowing of a pasture paddock, it is important that producers assess weed populations prior to, and after, the management tool is used. This can be done using a simple square quadrat and counting the number of weed plants within the quadrat. This is done a specified number of times in a defined area, that has been permanently marked for future reference. Producers can then return to that area in the following season and assess the impact of their management on the weed population. For species such as annual ryegrass, changes in populations occur after one year of management. For perennial species it may take successive seasons to observe changes in the weed population.

## VINEYARD WEED MANAGEMENT

Organic viticulture, like other organic industries, entails the production of high quality food products from agricultural systems that rely upon more natural approaches to crop nutrition, pest and disease control, and management of natural resources such as soil. The principles of organic viticulture include:

— the development and maintenance of good soil structure, fertility and biological activity, and
— the exclusion of most viticultural chemicals.

Weed management is one of the most challenging issues faced by grape growers adhering to these principles. Managers of organic vineyards need to achieve effective control of weeds without using synthetic herbicides and ideally with minimal disruption to the soil.

They meet this challenge by selecting from a wide range of acceptable techniques to develop weed management programs appropriate to their particular situation.

**Weeds in Viticulture**

There are many reasons why certain plants are, or become, weeds in vineyards. Plants are commonly defined as weeds if they exhibit one or more of the following characteristics:

— compete with vines for water (most plants);
— compete with vines for nutrients (most plants, especially non-legumes);
— interfere with water distribution from low-level sprinklers (most plants);
— harbour grape pests (most broadleaf plants host lightbrown apple moth);
— produce spiky seeds that may injure workers or contaminate produce (eg. caltrop or puncture vine: Tribulus terrestris);
— interfere with harvest or other vineyard operations (any tall plants);

— compete with more desirable cover crops (eg suppression of clover by capeweed: Arctotheca calendula);
— contaminate export tablegrape bunches with quarantinable seeds (eg. some thistles);
— affect vineyard aesthetics.
— have the potential to invade neighbouring properties where they can exert their undesirable qualities

**Classification of Weeds**

Before developing a weed management program, it is important that organic viticulturists consider their definition of weeds carefully. In many situations, considerable time, energy, money and other resources are spent combating 'weeds' that are in fact causing no physical or economic harm to the vineyard enterprise. As with the management of any pest, the use of resources to control vineyard weeds is only warranted when those weeds cause, or are very likely to cause, economic or other loss to the vineyard or neighbouring properties.

Volunteer plants in vineyards are often defined as 'weeds' simply because they were not deliberately planted as part of the grapevine/cover crop system. However, that definition alone does not justify efforts to control those plants. To determine their real weed status, the characteristics of volunteer species need to be considered, in relation to the management objectives of the vineyard and those of your neighbours. When these volunteer species are divided into 'real weeds' and 'other plants', the grape grower's job becomes easier.

***Weeds are not always weeds***

The real and perceived status of plants as 'weeds' can vary significantly between vineyards and regions because of differences in soil, climate, vineyard age, management practices, end use of the crop and the philosophy of individual grape growers. For example, grasses are often encouraged in cool climate vineyards on fertile soils where their competition for nitrogen is used to reduce excess vine vigour. On naturally poor soils of some hot, dry regions however, summer grasses tend to be discouraged, because they compete for water and nutrients – major issues for some regions. Summer ground cover does however help reduce vineyard temperatures and dust. Many agricultural weeds have been defined as such because of their negative impact in pasture and grain cropping systems, the dominant form of agriculture in Australia. Their impact, or role, in vineyards can be quite different.

Because of these differences, it is important that grape growers assess the weed status of plants in their own particular situation. However, consideration of potential weed spread to neighbouring properties should always be taken into account. Your weed management should be prioritised by: 1. Impacts on your vineyard, and 2. Potential impacts on your neighbours.

***Beneficial 'weeds'***

Some 'weeds' provide benefits of value to organic vineyard systems, and these benefits need to be weighed against any negative characteristics of the weeds. Some benefits of weeds are listed below:

— *Soil protection* - Most plants protect otherwise bare soil from the effects of sun, wind and rain.

— *Food source for beneficial insects* - Many flowering plants produce pollen or nectar that is a useful food source for beneficial insects such as Trichogramma (a wasp parasite of lightbrown apple moth), hover flies and lacewing insects (predators of mealybugs, scales and other vine pests) and predatory mites (predators of pest mites on grapevines). Other plants, like thistles, support aphids which provide a food source for beneficial insects and help maintain populations of those beneficial species in the vineyard.

— *Nutrient recycling* - Deep-rooted plants play a useful role in nutrient recycling by absorbing nutrients from lower in the soil profile and redistributing them when the plants decompose.

— *Weed suppression* - heavy growth of some 'weeds' like grasses, can suppress the establishment and growth of less desirable weeds such as the spikyseeded three-cornered jack.

***Pre-planting weed management***

Managing weeds before vines are planted, helps to reduce the weeds' competitive pressure during the establishment phase of the new vines. It also helps to reduce the risk of vine damage resulting from post-planting weed control practices such as cultivation.

Pre-planting weed control is especially important where the ground is infested with perennial weeds like couch grass (*Cynodon spp*), Johnson grass (*Sorghum halepense*) and nutgrass (*Cyperus rotundus*). Weeds such as these, established amongst vines, are much more difficult to manage than those growing in open ground.

***Established Vineyards***

Good weed management is critical for newly planted vineyards, because of the relative inability of new vines to compete for light, water and nutrients. Vines are most susceptible to competition from weeds during their first three to four years of growth. Inadequate weed management during this period can significantly delay the early growth and profitability of new vineyards. After this time, the vines' larger root systems permit them to compete with weeds more strongly. Shade from the vine canopy also helps suppress weed growth under established vines.

## INTEGRATED WEED MANAGEMENT

Integrated weed management (IWM) is an approach to weed management that helps growers achieve the most effective weed management outcomes in the short and long-term. Like IPM (integrated pest management), successful IWM relies upon some basic principles that include:

— correct identification of weeds
 — know what you are dealing with;
— some understanding of weed ecology
 — know why the weeds grow where they do, and how they get there;
— appropriate vineyard design
 — eg: choice of irrigation system can influence weed development and weed management approaches;
— choice of appropriate management methods
 — use effective techniques that minimise negative impacts on the vineyard environment (eg. soil);
— monitoring
 — know where weeds are a problem and how effective the control strategies have been.

Rather than relying upon one or two methods of weed control, IWM makes use of any appropriate methods to deliver the best outcome. One reason to integrate several weed control methods into a weed management program is that generally, no one method can control all weeds, costeffectively, all the time. This is because, for example,

— some weeds are easier to control than others;
— some weeds are annual and some perennial;
— some are spread by cultivation, others by wind;
— some are avoided by grazing animals;
— some are very competitive against cover crops.

By combining weed control methods, grape growers can reduce the risk of particular weeds dominating the vineyard and causing major problems.

### Total Property Weed Management

IWM programs should take the whole property into account. This helps minimise the risk of the actual vineyard area being continually reinfested by weeds carried by wind or on machinery from less-intensively managed areas such as roadsides, headlands, irrigation channels and habitat reserves.

Also, weed management needs to be integrated with overall vineyard management. Weeds, pests, diseases, soil, water and vines are all interrelated aspects of the vineyard system.

#### *Realistic objectives*

Weed management objectives need to be realistic and achievable if the programs are to have any chance of implementation, let alone success. For example, where curled dock (*Rumex crispus*) is considered undesirable, an objective to eradicate it from a vineyard within the short to medium term may be unrealistic, given that dock seed can remain viable in the soil for up to 80 years. A more realistic *objective would be to prevent further seed set of that weed.*

Like the actual definition of weeds, a grower's objectives and motivation in weed management should be considered carefully.

### Weed Management Techniques

#### *Quarantine*

Quarantine is used to minimise the introduction of weeds into the vineyard, and their spread between vineyard blocks. As a preventative measure, quarantine is very costeffective, compared to the resource input required to suppress introduced weeds over the long-term. This approach also reduces the risk of undesirable chemical residues entering organic vineyards. When using the quarantine approach, vineyard managers:

— restrict the movement of machinery, people and livestock onto and across the property, especially those coming from high-risk, weed-infested sites;
— clean machinery/equipment before it enters the vineyard, to remove weed seeds, bulbs, roots etc;
— take care in selecting clean sources of potted vines, mulch, raw materials for compost, or other materials that may be weed-infested;
— develop buffers such as windbreaks and surface water diversion, against wind and water-borne seed from adjacent properties.

### *Hygiene*

Hygiene is a component of quarantine, and involves additional elements that reduce the spread of weeds within a vineyard. As well as the points listed above, vineyard managers should:

— prevent weed seed set wherever possible;
— avoid activity that spreads weeds vegetatively, eg. avoid cultivating through areas infested with nutgrass, couch grass or other perennial weeds;
— destroy seed of noxious weeds by burning, deep burial etc. Grazing is not always a reliable technique, as the seed of many weeds germinates after passing through grazing animals.

### *Monitoring*

Monitoring is an important aspect of weed management. It helps grape growers to identify a site's major weed issues and hotspots, select appropriate management techniques, and determine the best timing of treatments (eg. soon after germination to prevent weed establishment or at the start of flowering to prevent seed set). Monitoring also reveals the effectiveness of weed control techniques. This is necessary for fine-tuning of weed management programs. For monitoring to be of value, it is important that weeds are identified correctly. Guides are available to help with this. Weed surveys and mapping are also a useful way to document a vineyard's weed status for future reference. Surveys should be carried out at times appropriate to the particular vineyard situation, eg. on drip-irrigated sites a week or two after significant rainfall events, to detect widespread germination of summer weeds, or after autumn breaking rains for winter weed germination.

*Competition*

Weed growth can be suppressed by competition from other plants for water, light, nutrients and space. A simple weed risk formula applies to organic vineyards like most other agricultural systems:

Bare soil = Weed problem!

In most ecosystems with adequate water resources, bare soil is unnatural, a sign of imbalance or disturbance. To counter this, many plants are adapted to colonise and cover bare soil very effectively. The result is that grape growers who attempt to maintain their system in a state of imbalance, ie. keep their vineyard soil bare, generally find themselves battling constantly against these colonising plants, or 'weeds'.

In contrast, many weeds are poor competitors, and their establishment, growth or seed production can be greatly reduced by a reasonable level of competition from more desirable plants. Wherever possible, growers should design bare soil out of their vineyard management system.

Cover crops and green manures are commonly used in organic viticulture to improve soil structure and fertility, and out-compete weeds in the inter-row area. Weed pressure is also reduced by the cultivation used to prepare seed-beds for these crops. The choice of appropriate crops will depend upon local climate, soil conditions and water availability. Mixtures of cereals and legumes are often the most competitive cover/green manure crop, but local expertise and experience should be used in choosing the right crop for each situation.

For effective weed suppression, green manure and cover crops need to be managed well so they cover the soil as quickly as possible to smother emerging weeds. Particular attention should be given to irrigation, nutrition and pre-and post-planting weed management, with such an implement as the Striegel, a flexible-tined finger harrow.

Another aspect of competition is that of allelopathy, which involves the suppression of seed germination or plant growth by chemical substances produced by another plant. These substances may be leached from the foliage or secreted from the roots of the growing plant, or released from the plant when it dies and decomposes. Many plants produce allelopathic substances. Of the potential cover/ green manure crops for use in organic vineyards, barley and cereal rye are known to suppress numerous weed species and are considered to have the greatest allelopathic potential.

### *Mowing*

Mowing is often used by organic grape growers to keep cover crops and weeds to a manageable height. It is a relatively fast operation that causes minimal soil disturbance, although soil compaction may become an issue where mowing is frequent.

The timing and height of mowing are important and can be manipulated to suit the grape grower's objectives. Mowing during or before flowering of weeds will prevent seed production. High mowing (eg. 20cm) will reduce or prevent seed set in medium to tall weeds while allowing useful species, including non-grasses, to regenerate. Low mowing tends to favour grasses over many other desirable plants, because the regeneration point for grasses is low. For the same reason, low mowing is not effective against weeds with very low crowns, like capeweed and caltrop.

Mowing of cover crops or resident vegetation can be used to produce mulch material for weed suppression in the inter-row or vine-row areas. If mulch is desired, slasher mowers are preferable to mulchers (eg. flail mowers). The coarser material produced by slashers will break down more slowly and have a longer effective life as a mulch.

Growers should consider a program of slashing alternate rows, allowing the uncut rows to provide habitat and food sources for beneficial insects in the vineyard.

### *Grazing*

Grazing has considerable potential for weed management in vineyards, and is in keeping with organic certification standards which encourage the incorporation of livestock into organic farming systems. In vineyards, sheep and geese are the most likely candidates because of their size and relative ease of management. A side benefit of grazing is that useful nutrient recycling occurs through conversion of weeds into manure. Well-managed grazing livestock can also provide a useful income source.

Grazing needs to be managed to avoid damage to vine shoots. In well-established Australian organic vineyards that use winter grazing for weed control, the sheep are removed from the vineyard before budburst to prevent them from eating the new vine shoots.

Grape growers using grazing for weed control need to address several management issues including:

— a need for temporary fencing or a suitable alternative to control stock movement and avoid over-concentration in specific areas;

— protection from predators (eg. dogs, foxes & eagles);
— selective grazing (eg. geese prefer grass) that can lead to dominance of specific weeds.

*Heat*

Thermal weeding uses heat in various forms (radiant, flame, steam, hot water) to kill plant tissue. This technique has been researched and used for weed control for many years, but is generally economically effective only against small seedlings. For this reason, thermal weeding is most commonly used on row-crop seedbeds, to kill the new flush of weed seedlings before the crop germinates. It has however also been used in vineyards and orchards.

A benefit of thermal weeding is that it does not disturb the soil. Organisms below the soil surface are generally not affected, as the heat does not penetrate far into the soil. A disadvantage of thermal weeding is its high energy cost - it is usually based on fossil fuel such as diesel or bottled gas. It is also a relatively slow operation to achieve a good weed kill. Also, flame weeders may create a fire risk, especially where there is dry ground cover or mulch. Research into thermal weeding continues, and new advances in technology will help make it a more economical and practical weed management option for organic grape growers.

*Rolling*

Weeds and cover crops may be flattened with a roller instead of being mown. Rolling leaves the plants largely intact, so they break down more slowly and create a longer-lasting mulch layer.

*Mulch*

Mulch inhibits weed growth and germination of weed seeds by blocking sunlight and light rainfall from reaching the soil surface. It also forms a physical barrier to weed growth. Almost any organic material can be used as mulch, with straw being commonly used in vineyards. Grape marc and composted green municipal waste is also used. Other materials include shredded paper, wet paper pulp (sprayed on) and jute matting.

The correct depth of mulch is important. Weeds will not be suppressed effectively if the mulch is too thin, and overly thick mulch can shed sprinkler water and rainfall away from the vine row. Some local experimentation may be required to determine the best mulch depth for each situation, but as a general guide, approximately 20cm of straw mulch is recommended, while only 7.5cm – 10cm of composted green waste mulch is required.

Mulching is commonly used for weed control in the vine row and is particularly useful on stony soils where undervine undervine cultivation is difficult or impossible. Good reasons for managing in-row weeds with organic mulch include:

— some mulches (eg. partly composted green waste) are a useful source of nutrients;
— the mulch adds organic matter to the soil;
— vine roots concentrated directly under the vine are protected from excessive heat;
— mulch greatly reduces evaporation, thus conserving soil moisture;
— mulching helps avoid vine damage from cultivation;
— vine trunks, sprinklers and irrigation tubing make cultivation and mowing difficult in the vine row.

Mulches do have some disadvantages that include:

— partly decomposed mulch may create an ideal seedbed for weeds;
— mulch may harbour pests such as snails;

— it can be difficult to manage weeds at the mulch/interrow interface without disturbing the mulch (steam weeding is an option).

The need for reapplication of organic mulches, possibly annually, has to be considered when costing this method of weed management. Also, as most organic mulch material is bulky, the cost of transport and handling needs to be considered.

Another consideration relating to mulch materials is their influence on soil microbial communities. 'Soft' materials such as straw or slashed cover crops are likely to favour bacterial decomposers, while fungi are more likely to be encouraged by 'hard' materials such as chipped wood. Growers who aim to fine-tune their soil management may need to take this into account.

Mulch costs may be reduced by growing mulch material on-site as inter-row green manures or cover crops. The mowing/mulching operation may be performed in a single pass with a side-throw mower that delivers the material directly to the under-vine bank.

This is a good approach for organic vineyards, as it reduces the need for external inputs. It also reduces the risk of bringing unwanted weeds into the vineyard, in contaminated mulch material. Synthetic weed mat effectively suppresses some weeds, and a temporary mulch of black plastic may be used to kill patches of difficult weeds including couch grass, by starving them of water and sunlight.

This technique does not need the hot conditions required for solarisation, but has to be applied over months rather than weeks. Disposal of these materials is an environmental issue that organic growers would need to consider. Also, their use may be restricted by organic certification organisations.

### *Biological control*

Like most plants, weeds are subject to attack by their natural enemies including root, stem and seed-eating insects, mites and diseases. In some cases these organisms are officially imported, and, after considerable research, released for the control of introduced noxious weeds. There have been some spectacular successes in the biological control of major weeds in Australia, but most common vineyard weeds have not warranted this level of attention. Weeds that are, or have been, the target of biological control in Victoria are listed in Table 1.

*Table 1. Weeds targeted for biological control in Victoria.*

| | |
|---|---|
| Blackberry | *Rubus fruticosus agg.* |
| Boneseed | *Chrysanthemoides monilifera* |
| Bridal creeper | *Asparagus asparagoides* |
| Cape broom | *Genista monspessulana* |
| Chilean needle-grass | *Nassella neesiana* |
| Common heliotrope | *Heliotropium europaeum* |
| Cord grass | *Spartina spp.* |
| Dock | *Rumex spp.* |
| English broom | *Cytisus scoparius* |
| Flax leafed broom | *Genista linifolia* |
| Gorse | *Ulex europaeus* |
| Horehound | *Marrubium vulgare* |
| Illyrian and Scotch thistles | *Onopordum illyricum & O. acanthium* |
| Paterson's curse | *Echium plantagineum* |
| Prickly pear | *Opuntia stricta* |
| Ragwort | *Senecio jacobaea* |
| Serrated tussock | *Nassella trichotoma* |
| Skeleton weed | *Chondrilla juncea* |
| Slender thistle | *Carduus pycnocephalus &* |
| *Carduus tenuiflorus* | |
| Spear thistle | *Cirsium vulgare* |

| | |
|---|---|
| Spiny emex | *Emex australis & Emex spinosa* |
| St John's wort | *Hypericum perforatum* |
| Tiger pear | *Opuntia aurantiaca* |
| Tutsan | *Hypericum androsaemum* |
| Variegated thistle | *Silybum marianum* |

Biological control programs are generally operated by the Government rather than by individual land holders. Anyone seeking information on these programs in Victoria should contact the Victorian Department of Primary Industries Frankston.

*Soil modification*

Different plants are adapted to different soil conditions like high or low pH, salinity, fertility, moisture content, compaction and so on. Because of this, the presence, dominance or quality of growth of particular plants, including weeds, may indicate to some degree, the underlying soil conditions.

Awareness of this relationship between weeds and soil gives vineyard managers some opportunity to address their weed issues by modifying their soil conditions. In the long-term, this may be more cost-effective than tackling the weeds directly.

Weeds are not always 100% reliable as indicators of specific soil conditions because, for example:

— some weed species are tolerant of a wide range of soil conditions;

— some are sensitive to a number of environmental conditions;

— some species develop localised 'ecotypes' that differ in their adaptation to specific conditions.

For this reason, care should be exercised when using weeds as soil indicators. Communities of weeds, rather than individual species, are likely to give more reliable indications of soil conditions.

Numerous sources exist for information on 'indicator plants' but not all are reliable. Growers should compare several information sources and combine them with their own observations as a basis for interpretation of the soil conditions indicated by their weed communities.

*Mechanical cultivation*

Cultivation destroys or buries weeds, disrupting their growth and preventing seed set. Most forms of cultivation are very disruptive to soil structure and to soil organisms. Cultivation can also spread seeds, tubers and rhizomes of noxious weeds, is relatively energy-expensive and increases the risk of soil compaction. Despite these costs, cultivation is common practice in organic viticulture. Standard tools such as disc or chisel/tine ploughs are generally used to cultivate the vineyard inter-row area.

These tools are mounted on manually or automaticallyoperated retractable arms, so they can be moved around vines and irrigation risers. With this equipment, experienced operators can clean vine rows of weeds relatively quickly, with little if any damage to the vines. Some growers use these or other tools to control annual weeds with the scrape-off/throw-on strategy. Soil is repeatedly shuffled on and off the under-vine area to destroy weeds where the soil is removed, and smother those where the soil is dumped.

In hot dry climates, under bare soil, grapevine roots usually avoid the top 100mm or so of soil, so the root system will be largely unaffected by shallow cultivation. In cooler regions, or where mulch has been applied, shallow roots will develop and these will be destroyed by undervine cultivation.

*Solarisation*

Solarisation involves the destruction of plants and seeds by the solar heat trapped when soil is covered by clear plastic sheeting during hot weather. While this technique is more commonly used to kill disease organisms in the soil, it is also effective against many weeds and their seeds. However, it is generally not effective against weeds with deep tubers or rhizomes, like Johnson grass.

Effective solarisation requires good movement of heat into the soil. This is aided by high levels of soil moisture, so, if necessary, the soil should be irrigated before the plastic is laid down. Air gaps between the plastic and soil should be minimised as they inhibit heat transfer into the soil. To prevent the plastic being lifted by wind, and to reduce the leakage of heat, the edges of the sheet should be buried.

Solarisation for at least four weeks is likely to be needed for effective weed kill. This will vary according to the quality of plastic sheet used, the prevailing temperatures and weed species present. Winter weeds are likely to be less tolerant of high temperatures than summer weeds, and therefore more strongly affected by solarisation. To avoid reinfestation of treated areas with weeds, the soil should be mulched or sown to a green manure/cover crop as soon as possible after removal of the plastic. Avoid bare soil!

*Hand cultivation*

Organic grape growers often do some hand weeding to remove weeds from around young vines or weeds that are too close to vine trunks to be removed by undervine cultivators or mowers. Although hand weeding is timeconsuming, and therefore expensive, it is effective and gives growers an opportunity to observe their soil, vines and vineyard closely.

*Chemical*

Organic knock-down herbicides based on plant extracts like pine oil are available commercially. These herbicides disrupt the cuticle of plants and cause them to desiccate. Even though these herbicides are organic, they still involve the application of chemical substances to the vineyard ecosystem, and should be used with care, as the effects of their long-term use are not yet known. The use of organic herbicides to replace synthetic herbicides also diverges from the spirit of the organic approach, which involves considering and managing the vineyard differently, rather than simply substituting allowed inputs for prohibited inputs.

The organic herbicides currently available need to be applied at relatively high concentrations and seem to work well only against small seedlings. Because of this, their cost-effectiveness should be assessed carefully before organic grape growers incorporate them into weed management programs.

Growers considering the use of organic herbicides should first obtain details of any permitted products from the relevant organic certification organisation.

### *Timing*

In some situations, the timing of weed control treatments in relation to the weeds' developmental stage or weather conditions is critical - a few days can make the difference between success and failure. Weed management actions should be prioritised, scheduled and performed at the optimum timing for best results.

If management of a specific weed relies upon the prevention of seed set, then control treatments must be given adequate priority and applied before, or during flowering. Whenever weeds are allowed to produce a new generation of seed, the problems caused by those weeds will be perpetuated, as will the need for ongoing control. For example, a single plant of blackberry nightshade (*Solanum nigrum*) can produce over 170,000 seeds.

Similarly, cultivations to disrupt weed seedlings or to chop the rhizomes of perennial weeds will be more effective if followed by hot, dry weather rather than cool, wet weather. Priority should therefore be given to timing such cultivation according to the prevailing weather conditions.

## WEED CONTROL IN ORGANIC CARROTS

The production of good quality certified organic carrots is generally seen as difficult. Many growers believe that cost effective, adequate weed control without the use of herbicides is too hard to achieve. However, this is essential for mechanical harvesting and a good yield. The good news is that effective implements exist to achieve a good seedbed and to control weeds for a weed-free carrot crop.

With carrot seed being small (averaging 800 seeds/g), it is important for good crop establishment to sow into a deep, weed-free seedbed of good structure. The slow germination of carrot seed gives weeds a head start. They grow faster than carrot seedlings and compete strongly for nutrients and sunlight in the early stages of crop growth. Unchecked, weed growth during this critical time has the most detrimental impact on yield. Thus, to obtain a good yield and ease of mechanical harvesting, it is essential to control emerging weeds from the time of seedbed preparation until the carrot tops have grown tall enough to shade out weeds. There is a range of soil tillage, bedforming and mechanical weeding implements at the organic grower's disposal to achieve this.

### Mechanical Tillage and Weeding Implements

Growers use different types of implements at different times of carrot crop production:

*Pre-sowing:*

— Various types of tillage implements

— Various types of bedformers
— Weeding implements for first flush of weeds

*After sowing – pre-germination of carrots:*

— Flame or steam weeder
— Harrow for blind weeding

*After germination of carrots:*

— Various types of tine cultivators
— Brush weeder
— Row flame weeder
— Hand hoe

**Steps to Implementation**

*Pre-sowing*

*Tillage implements*

Either a moldboard plough, disc plough or disc harrow is commonly used in Victoria as the pre-planting tillage implement for carrot seedbed preparation. The mouldboard plough inverts the soil, buries trash, weeds and weed seeds. Its operation is very energy efficient, but needs high draft power. The disc plough also penetrates the soil well and inverts the soil. It cuts and covers weeds and trash, and breaks up clods. However, it can also be severe on soil structure. Disc harrows cut up trash and weeds well, but leave some of this trash on or close to the soil surface, which may interfere with carrot crop establishment. Disc harrows can also damage soil structure.

The blade plough is an implement that cuts weeds below the soil surface and leaves them lying on the top of the soil while largely preserving soil structure. It leaves a level bed ready for a final pass with a rotary hoe type bedformer.

*Bedformers*

Tillage with a plough or disc harrow is usually followed by a bedforming operation. A simple bedformer is a frame carried on the 3-point linkage. It pushes the soil from the wheel tracks into the centre of the bed and flattens the soil to create a uniform, level, full width bed of about 20 cm height.

Rotary hoeing and bedforming in one operation is a time saving alternative to ploughing or discing followed by bedforming. The implement used is a power-take-off driven rotary tiller with curved blades on a horizontal axle, embedded into a bedforming frame. Its advantage is that with one pass, trash and weeds are destroyed extremely well and beds are formed ready for sowing. Disadvantages include the potential for soil pulverisation, destruction of earthworms and other large soil fauna, and the high tractor power demand and energy use involved in operating this implement.

Fields vary in the types of weeds and size of weed seedbank in the soil. Sometimes it might be advantageous to irrigate the prepared seedbed to initiate the germination of a first flush of weeds. These weeds need to be killed with minimal soil disturbance, so as to prevent bringing fresh seeds to the surface where they will germinate. A shallow cultivation of 2.5 cm to 5 cm depth can be done with a rigid or finger-tine harrow, or a vibrating cutting bar. The latter implement requires a trash-free seedbed. Usually a roller is attached to either weeding implement to crush clods and leave a smooth, flat seedbed for the following crop sowing operation.

*After sowing, but pre-emergence of carrots*

Carrots usually take longer to germinate than most weeds. This allows the grower to control a flush of weeds after sowing, with either blind harrowing or a flame weeding or a steam weeding equipment, before the carrots emerge. This operation greatly reduces the difficulty of achieving effective weed control after carrot crop emergence. A 3-point linkage flame weeder has several LPG-fuelled burners that cover the whole bed. A steam weeder creates steam and blows it over the weeds. It is safer to use in a dry environment, because there is no danger of setting dry plant material alight.

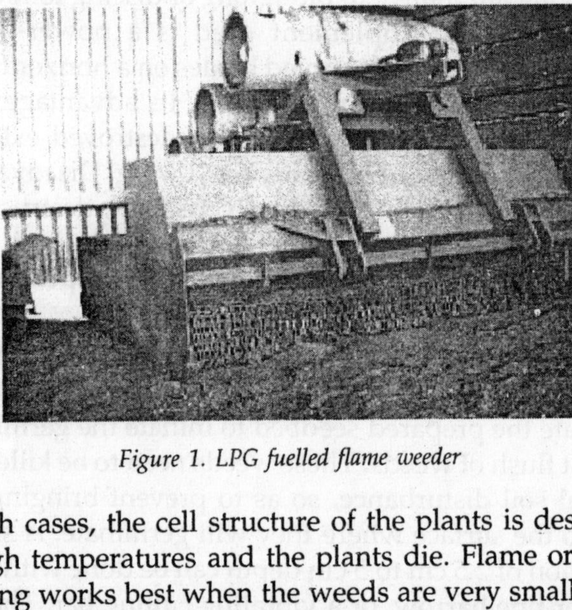

*Figure 1. LPG fuelled flame weeder*

In both cases, the cell structure of the plants is destroyed by high temperatures and the plants die. Flame or steam weeding works best when the weeds are very small. Their great advantage is the control of weeds between the carrots (intra-row), which greatly reduces subsequent hand hoeing. The heat of the flame or steam applied to the soil neither damages the carrot seeds in the ground nor causes a significant reduction in soil microbial biomass.

*After emergence of carrots*

Careful soil cultivation between the carrot rows (inter-row) after the carrot seedlings have emerged, controls weeds and also activates nutrient mineralisation and release. The soil disturbance caused by cultivation exposes locked up organic matter and minerals to oxygen, water and microorganisms for decomposition. This may boost crop growth. Weeding implements for this task are best mounted 'underbelly', i.e. underneath the tractor between the front and rear wheels for accurate steering, or equipped with a mechanism such as 'hugboards' or 'bedlockers' that guide the implement along the raised seedbed. These steering mechanisms keep the cultivator over the centre of the bed, and with carefully adjusted tine spacings crop damage is avoided.

*Tine cultivators*

The traditional and still most common types of weeding implements are tine cultivators. Different shapes and arrangements of tines and blades are used with respect to soil type, stage of weed growth and grower preference. Blade-tine cultivators have a cutting, uprooting and weed covering action and usually give good weed control. However, they tend to leave a furrow and create some soil compaction. A finger-tine harrow, consisting of flexible, spring loaded, 5mm diameter round rods on a flexible frame, is soft on soil structure and avoids the drawbacks of the blade-tine cultivator.

However, as it has no cutting action, the finger-tine harrow only works well with small weeds. Also, intra-row weeds are not controlled, as insufficient soil is thrown intc the carrot rows to cover emerging weed seedlings.

*Rotary-tine cultivator*

The Weed-fix® is a row-crop tine cultivator consisting of small, flattened rods as tines arranged in circles. These hydraulically driven groups of tines are positioned over the inter-row spaces and rotate to cultivate the soil and uproot the weeds. Working width and depth can easily be adjusted. Adjustable boards to the sides of the rotating tines allow variable amounts of soil to fall into the crop row for intra-row weed control.

*Figure 2. Rotary tine cultivator (Weedfix ®)*

Compared to the other weeding implements discussed here, the Weedfix® is the most adjustable weeding implement that can be used effectively at a greater range of weed and crop growth stages. Table 2 shows the weeding results with a goosefoot-tine cultivator, a finger-tine harrow and a rotary-tine cultivator (Weed-fix®).

Weed control was equally effective to about 80% of weeds killed with all three implements. However, the number of carrot seedlings either covered with soil by the goosefoottine cultivator or weeded out with the finger-tine harrow were considerable.

*Table 2. Reduction in numbers of carrot seedlings and weeds by three different weeding implements*

| *Goosefoot tine cultivator* | | *Finger-tine harrow* | | *Weed-fix®* | |
|---|---|---|---|---|---|
| Carrots | Weeds | Carrots | Weeds | Carrots | Weeds |
| 67% | 83% | 33% | 82% | 0% | 80% |

This amounted to a substantial yield loss. The performance of the finger-tine harrow would be improved greatly, if a bed-guiding mechanism was attached to the implement.

*Brush weeder*

An inter-row brush weeder consists of rotating nylon brushes that brush 3-5 cm deep over the topsoil and uproot and destroy small weeds. The brush weeder pictured has vertical nylon brushes mounted on a horizontal axle rotating in the driving direction.

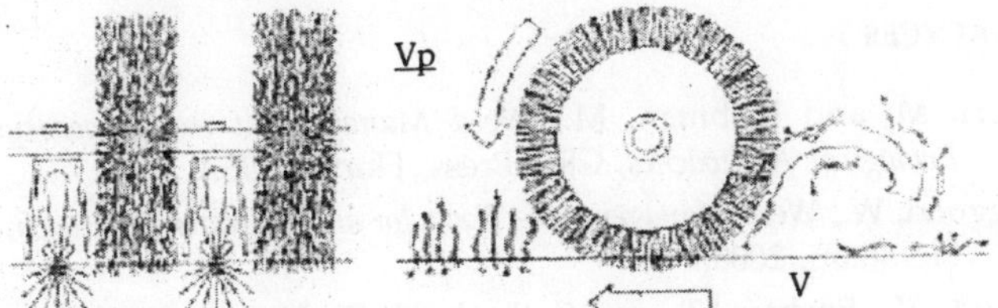


*Figure 3. Weeding action of vertical brush*

This type of brush weeder is designed to weed closely to the crop to minimise hand weeding, but the brushes do not throw soil into the carrot row for weed control. Another type of brush weeder has horizontal brushes on vertical axles. The direction of the brush rotation can be selected to either throw soil into the crop row or not.

*Inter-row thermal weeders*

An inter-row flame or steam weeder that is used when the weeds are very small eliminates any inter-row soil disturbance and thus reduces the potential for bringing new weed seeds to the surface. However, no soil at all is thrown intra-row, i.e. all intra-row weeding must be done by hand.

*Hand hoeing*

None of the implements used for carrot weeding eliminate hand hoeing between the carrots altogether. Several types of handheld hoes are available. Most commonly used are various shapes of steel cutting blades on a long wooden handle. A different design is a 'wire weeder', which consists of a round steel rod forming a loop at the end of a wooden handle to which it is attached. The wire weeder is soft on soil structure and crop plants. However, it works best in a light soil and when the weeds are small, as the rod is not sharp enough to cut weeds off.

## REFERENCES

Altieri, M. and Liebman, M., *Weed Management in Agroecosystems: Ecological Approaches*, CRC Press, Florida, USA, 1988.

Bedggood, W., *Weed Management – Tools for an Integrated Approach*, NRE, Horsham, 2000.

Burnett, V., Enshaw, T. and Sutherland, S., *Non-chemical Options for Integrated Weed Management in Grain Production*, Victorian Department of Primary Industries, 2004.

Buckerfield, J. and Webster, K., "Vineyard trials show value of mulches - organic matter for *management of young vines*, Australian Grapegrower and Winemaker #441, October 2000.

Buckerfield, J. *Compost mulch-good and evil*, Acres Australia, April/May 2002.

Chase, C.A., *Soilborne Plant Pathogen and Pest Management with Soil Solarization*, University of Florida, Institute of Food & Agricultural Sciences, 2002.

Cousens, R. and Mortimer, M., *Dynamics of weed populations*, Cambridge University Press, Cambridge, UK, 1995.

Sindel, Brian M. Ed., *Australian Weed Management Systems*, R.G. and F.J. Richardson, Melbourne, Australia, 2000.

Tenni, A. and Faithfull, I., "How to manage weeds", *NRE Landcare Notes*, 1998.

# 9

# Organic Agriculture and Biomass Production

Bioenergy is becoming an ever more important option in climate change mitigation policies. For a sustainability assessment, different types of biomass have to be distinguished. Biomass from relatively sustainable forest use practises or perennial bioenergy crops on marginal or highly erodible land perform differently than high-yield energy crops with annual harvest, for example.

As in food crop production, large-scale monoculture production of biomass is likely to be particularly unsustainable. But even employing 'marginal land' for biomass production for energy use may well result in averse effects for local communities if not assessed with a true understanding for their local livelihood strategies.

Besides biomass specifically grown or harvested for energy use, bioenergy can be produced from biomass waste from agricultural production such as bagasse, rice husk, fruit shells from palm oil production or other agricultural residues. Using these resources for energy production is more sustainable than having them decay without use and with considerable methane emissions. The sustainability of burning these resources for energy production can however be questioned in the context of their potential use as fertiliser under sustainable agriculture practises.

This chapter attempts to contribute to the discussion on sustainability of bioenergy production on a general level by combining knowledge from different disciplines, mainly from agricultural sciences and (bio-)energy systems. It relates a potentially huge production of biomass for energy use or a potentially huge demand for agricultural waste to the sustainability of current agricultural production systems and to alternative agricultural production practises. I formulate the hypothesis that an aim to provide both sustainably grown agricultural products and bioenergy, should they contribute significantly on a global scale, adds another sphere of potentially strong competition to the one on land and water between food and energy crops.

How to judge sustainability of bioenergy on a project level and for specific biofuels is a topic of ongoing discussion. Jürgens and Muller present several existing sets of sustainability criteria for bioenergy projects in comparison with sustainability indicators for general climate change mitigation projects and identify an important mismatch especially for issues related to the production of biomass. While present in many criteria sets for bioenergy, this topic is almost completely lacking in the general sets of criteria. But also in the bioenergy context, the sustainability discussion is mainly confined to specific situations and crops and does not relate to aggregate effects on a global level.

## AGRICULTURE, LAND USE, WATER SCARCITY AND BIOENERGY

The total global area currently used in agriculture (arable land and permanent crops) is about 1530 Mha. Potential arable but not yet used land for rain-fed agriculture is estimated to be about 2800 Mha, whereof 45% are covered with forest. However, much of this land is needed to preserve forest cover and for infrastructural development. Accessibility puts further constraints to any substantial expansion. In addition, these land reserves are distributed very unequally. More than half of them are located in seven countries only. Additional 200 Mha are available when irrigated in developing countries. How much of these total land reserves will be available for agriculture in 50 or 100 years is far from clear, however, because of losses in arable land due to soil degradation and water scarcity in the context of conventional agricultural systems, and as it is likely that climate change will on aggregate negatively affect agriculture and the suitability of land for farming in not temperate climate zones.

Azar assumes a maximum availability of 200 EJ/year from biomass as a reasonable estimate for the potential of bioenergy use. The review of 9 future bioenergy supply scenarios given in Bauen et al. shows that identify a potential of 100 - 250 EJ/year in 2050. The total range of these 9 studies investigated is somewhat larger though, from 100 to 450 EJ in 2050. Berndes et al. assess a larger but partly overlapping set of 17 bioenergy potential studies. They report a similar range, from 100 to 400 EJ/a in 2050. Hoogwijk et al. present a more detailed and recent assessment of the bioenergy potential in four IPCC-SRES land-use scenarios. They find a higher total potential with a range between 130 and 410 EJ/a in 2050 on abandoned agricultural land and an additional potential of 35 to 245 EJ/a in 2050 on the partly available other land area. In the model of Azar, the global population would be 10 bill people by the year 2100 and it is assumed that each global citizen would consume services corresponding to about the same amount of primary energy as the average OECD citizen today, i.e. corresponding to a per capita primary energy consumption of about 200 GJ/year. By 2100, half of it would be met by energy efficiency measures. In his model, the assumed bioenergy supply would then account for more than a fifth of global primary energy supply to be generated by biomass by 2080 (dropping to about a seventh by 2100, in absolute terms decreasing by about a fourth). Gross estimates indicate then that the corresponding energy crops would use about 500 Mha, i.e. an area equalling a third of the currently used area in agriculture. Per se, these 500 Mha may not seem that big a number in relation to the 3000 Mha from above. However, only a fraction of those 3000 Mha will realistically be available, the distribution of this soil resource is also important, and population growth over the next century will result in growing demand for food and land. This suggests a potential competition for land between food and energy crops.

Azar investigates the effects of such an expansion of land use due to biofuel demand on land rents and crop prices. This addresses the first shortcoming of existing studies on bioenergy availability as identified by Berndes et al.: "...the studies do not provide much insight into how the expanding bioenergy sector will interact with other land use". By the end of the century, food crop prices would reach levels three to five times as high as today.

Some rise in food prices from the currently very low levels would benefit the farmers. With too high prices, though, negative impacts could dominate, especially on the landless and the urban poor. In 2006, 850 mill. people were undernourished. Although a large part are smallholders and farmers or agricultural workers, it is not clear if they would benefit from higher land rents and food prices. The interactions are complex and effects are different for different groups. This is illustrated, for example, by some case studies in the context of increased land rents and crop prices due to increased demand for a particular crop. It is found that potential profits might well be captured by powerful large producers and lobbying groups. It is further expected that the already vulnerable part of the societies will be suffering from the direct effects of climate change itself, independently of a potentially increasing food-energy crop competition.

A second issue recognised as a potential problem for bioenergy and discussed in detail in Berndes is the interaction of increased biofuel production with the projected increased water scarcity in many developing countries and the corresponding increase in dependence from cereal imports ("virtual water"). While specific statements are difficult to provide, Berndes concludes that "...assessments of bioenergy potentials need to consider restrictions from competing demand for water resources." Like malnutrition, water scarcity is not a matter of total availability of supply but of distribution.

From now 450 mill people living in water scarce or stressed countries (i.e. not able to assure the minimal demand of 1700 $m^3$/cap/a for the population), this number is expected to rise to some 2800 mill. by 2025. It is likely that higher food prices will considerably affect these nations as a whole. Most of the nations importing cereals today are oil-rich and/or do not belong to the low-income countries as classified by the World Bank, but many countries that will become net cereal importers in the future are poor and hence unable to afford to purchase cereals. On the global scale, it is estimated that water scarcity will be the driving force of about 25-30% of the global cereal market.

Besides competition for water and land between food and bioenergy crop production, direct competition for biomass for different use is also an issue.

The Project Development Documents (PDD) of the Clean Development Mechanism (CDM) provide a number of examples. Bioenergy projects may constrain fuel supply for local industries such as brick makers. Such projects resulting in increased demand for biomass could also affect farm practises relying on composting crop residues, etc. and using them as fertiliser. From an economic point of view, this is no problem, as it is a local or regional readjustment to allocate the raw material to the most efficient use and thus reflects market mechanisms. These effects are thus not technical but pecuniary externalities only. Nevertheless, such changes in local or regional structure can potentially have adverse effects and counteract the general goal of poverty alleviation if the new activities do not offer or are accompanied by viable income alternatives for the population or by other measurements to avoid hardship. The competition on the crops itself, such as mentioned above for corn in Mexico/US, may serve as a drastic example of the short term adjustment problems that might occur. Competition for the biomass itself can arise in various contexts of alternative use.

## CONVENTIONAL AGRICULTURE

One main problem with the proposal of bioenergy as a valuable globally significant option to reduce greenhouse gas emissions is the neglection of the production process. There are detailed studies acknowledging this to a certain extent, pointing out the particular relevance of fossil fuel and fertiliser use in an ecological assessment of bioenergy production), but in many overview or outlook studies, bioenergy is, regarding availability, often seen somewhat similar to fossil fuel: It is assumed to basically be available, subject only to the availability of land and inputs of labour and capital. Thus the production process for biofuel in common economic models is not modelled in any more complex manner than oil or gas extraction and refining. However, as adequate the simple picture of extraction for general assessment of fossil fuels may be, as the production took place millions of years ago and does not matter anymore, it is not adequate for agricultural production.

Agricultural production has to take place annually now and in the future. To grow plants is a highly complex process that involves much more than only land, capital and work on such an aggregate level. This is evident when looking at how highly unsustainable conventional agriculture often is. Expansion of agricultural land is one of the most significant human alterations of the global environment and conventional agricultural production has often adverse effects on ecosystems. Problems regarding long-term sustainability arise on local, regional and global level.

Acknowledging its immense successes regarding crop yields and food security and that it relieved poverty for hundreds of million of people between 1965 and 1990, the green revolution (based on monocropping with high yield species, irrigation and increased use of chemical fertiliser, herbicide and pesticide input) also led to degradation and salinisation of soils, degradation and overuse of water bodies, loss in biodiversity - a list that could be prolonged. Besides these adverse environmental effects, it has to be stressed that a large part of the rural poor gained little from the green revolution and that poverty reduction through these techniques has slowed down. After forty years, this heritage has left a negative legacy in many countries.

It is unclear if this high intensity agriculture can be further sustained. A further large increase in these production techniques due to the increase in biomass demand for bioenergy would exacerbate this problem. There is some awareness of the potential difficulty to sustainably produce this large amount of biomass in the climate community. But in contrast to land competition, it has never really entered the discussion. Giampietro and Ulgiati are a notable exception arriving at dull perspectives for the sustainability of large-scale biofuel production, based on prospective land, water, labour and pesticide requirements and other impacts on society.

## SUSTAINABLE BIOENERGY PRODUCTION

To avoid the problems of conventional agriculture, biofuel would need to be produced sustainably. There are basically two ways - using waste biomass that would otherwise be unused and decay, or growing bioenergy crops sustainably.

There are several approaches of "sustainable agriculture". One option to sustainably grow biomass for energy use may be some forestry systems, where nutrient loss can be kept on a low level by on-site foliage and re-application of wood ash, for example. How sustainable these practises are in the very long-run, however, needs further investigation and paradigms of sustainable forestry and development of adequate sustainability indicators are still subject to discussions, especially in the light of its increasingly important role in the bioenergy context.

Using waste biomass for energy production may be less problematic at first sight, but I will argue that also this strategy faces incompatibilities if combined with organic agriculture. Using biomass waste for energy production eliminates the negative side effects of open biomass waste burning or deposition and generates useful heat and power as well. Examples are waste from rice (rice husk), sugar (bagasse) or palm oil (fruit shells) production. A large number of such projects are realised under the CDM and detailed information is available from the PDDs. As already mentioned in the introduction, the sustainability assessment of such projects is, however, no easy task.

Biogas production from organic material by capturing methane from anaerobic fermentation of dung and manure can be another option for sustainable bioenergy. Bath et al. present a study on dung-based biogas. Such projects are also realised under the CDM, e.g. the biogas projects on household level Begapalli and Biogas Nepal Activity I and II or industrial projects based on swine manure. The raw material is no waste anymore and the fermented material can be applied as a high quality fertiliser.

Although it is thus questionable if a restriction to truly sustainable sources could supply the amount of biomass needed in order to provide a significant share of global energy use, these energy sources can clearly be relevant on the farm or maybe at the community level.

## Organic Agriculture and Closed Cycle Resource Use

The principles of organic agriculture are, as stated by the umbrella organisation for the organic movement IFOAM (International Federation of Organic Agriculture Movements),

— the *Principle of health*: Organic Agriculture should sustain and enhance the health of soil, plant, animal, human and planet as one and indivisible;

— the *Principle of ecology*: Organic Agriculture should be based on living ecological systems and cycles, work with them, emulate them and help sustain them;

— the *Principle of fairness*: Organic Agriculture should build on relationships that ensure fairness with regard to the common environment and life opportunities; and

— the *Principle of care*: Organic Agriculture should be managed in a precautionary and responsible manner to protect the health and wellbeing of current and future generations and the environment.

or, as in a paraphrase by Eyhorn et al.:

"Principles and Aims of Organic Agriculture - A System Approach:

> Conventional farming puts its focus on achieving maximum yields of a specific crop. It is based on a rather simple understanding: crop yields are increased by nutrient inputs and they get reduced through pests, diseases and weeds, which therefore must be combated. Organic agriculture is a holistic way of farming: besides production of goods of high quality, an important aim is the conservation of the natural resources fertile soil, clean water and rich biodiversity. The art of organic farming is to make the best use of ecological principles and processes. Organic farmers can learn a lot from studying the interactions in natural ecosystems such as forests."

Organic agriculture is often criticised for achieving too low yields to feed the world. Drinkwater et al. report no significant differences between conventional and organic farming in yields for maize and Maeder et al. report 20% lower yield in organic farming on average, but with 30 to 50 % lower fertiliser and energy input, and almost no pesticides. See also FAO and Eyhorn and references therein for a general assessment and some socio-economic aspects. The studies on yields have predominantly been undertaken in temperate climate zones and there are hardly any studies on the performance of organic farming in the Global South. Based on a review of some case studies, Parrott and Marsden conclude that yields in organic farming systems in the South are not lower than for conventional systems. This is also the result of a study on cotton in Madhya Pradesh, India.

Organic agriculture focuses on nutrient cycles, soil protection, crop diversity and bio-control of pest and weeds in organic farms. These issues are closely interlinked. Nutrient cycles are closed with the help of composting, mulching, green manuring, crop rotation, etc. and nutrients exported from the farm with the sold produce (food, cotton, etc. - or biomass for energy use) need to be replaced in some way. Food export from the farm is only a part of the total biomass grown, whereas for bioenergy, it can approach close to 100%. This material is taken away and burned, which severely interferes with closed nutrient cycles.

Most plants can only take carbon and oxygen from the atmosphere, some species nitrogen as well (e.g. Legumes by symbiosis with microbes). But most nitrogen and all the other nutrients are supplied via the soil. Thus either the stock of nutrients in the soil is run down, which is clearly an unsustainable solution, or it is replaced either by natural or chemical fertilisers. For bioenergy production mainly the latter option is available and the possibility of sustainable production is therefore questionable. Ash recycling does not solve the problem as ash is a mineral fertiliser containing mainly kalium, calcium and trace elements. It has to be complemented by other fertilisers to deliver organic matter, phosphorous and nitrogen.

Clearly, bioenergy also builds on closed cycles - but only for carbon, while the limits for sustainability on farm level are rather set by a closed nitrogen cycle. For bioenergy, the carbon cycle also matters only on the most elementary level of carbon as a chemical element and as a part of the atmosphere in the form of the structurally very simple carbon dioxide. The closure of the cycle is achieved by requiring that as much $CO_2$ is to be taken from the atmosphere as is emitted by burning the fuel. In contrast, the cycles in organic agriculture are not closed on the level of chemical elements but rather of the more complex compounds contained in organic material. In consequence, soils under organic agriculture are usually higher in organic matter content, biological activity, and less vulnerable to erosion. Biodiversity is also higher and the crops can profit from root symbioses and exploit the soil better. These services cannot be delivered by inorganic chemical fertilisers focusing on nutrient input on the much lower complexity level of chemical elements. It thus seems to be impossible to burn biomass to a significant extent in the context of an organic farming system.

## Sustainability and Bioenergy

The approach of organic agriculture is significantly different from what seems to be seen as sustainable in parts of the bioenergy community - compare for example the recent volume in "Biomass and Bioenergy" on this topic. Organic farming is an explicit topic only in one article dealing mainly with possible on-farm use of bioenergy like short rotation coppice. However it claims one goal of organic farming to be to reduce non-renewable energy use and correspondingly focuses on this and less on the question of closed nutrient cycles.

## REFERENCES

Berndes G., "Bioenergy and Water—the implications of large-scale bioenergy production for water use and supply", *Global Environmental Change,* 2002.

Berndes G, Hoogwijk M, van den Broek R., "The Contribution of Biomass in the Future Global Energy Supply", *A Review of 17 Studies,* Biomass and Bioenergy, 2003.

Delucchi M.A., "Life Cycle Analysis of Biofuels", *Draft Manuscript,* Davis, CA: ITS, University of California at Davis, 2006.

Eyhorn F, Heeb M, Weidmann G., "IFOAM Training Manual for Organic Agriculture in the Tropics", International Federation of Organic Agriculture Movements (IFOAM), 2003.

Eyhorn F., "Organic Farming for Sustinable Livelihoods in Developing Countries? The Case of Cotton in India", vdf Hochschulverlag, ETH Zrich, Switzerland, 2007.

Richardson J., "Sustainable production systems for bioenergy: Impacts on forest resources and utilization of wood for energy", *Biomass and Bioenergy,* 2006.

Scott D.A, Dean T.J., "Energy Trade Offs Between Intensive Biomass Utilization Site Productivity Loss and Ameliorative Treatments in Loblolly Pine Plantations", *Biomass and Bioenergy,* 2006.

# 10

## Organic Farming and Climate Change

Climate change will dramatically alter global food production. The inequity in food supply between industrialised and developing countries is expected to increase, as the 40 poorest countries in the tropical and subtropical zones will suffer most, both from droughts and periodic floods.

Agriculture is not only affected by climate change but also contributes to it. Ten to twelve percent of global greenhouse gas emissions are due to human food production. In addition, intensive agriculture has led to deforestation, overgrazing and widespread use of practices that result in soil degradation. These changes in land use contribute considerably to global $CO_2$ emissions. Sustainable agriculture and food supply systems are thus more urgently needed than ever before. They must boost the capacity of agricultural production to adapt to more unpredictable and extreme weather conditions such as droughts and floods, reduce greenhouse gas emissions in primary food production and halt or reverse carbon losses in soils.

Organic agriculture is claimed to be the most sustainable approach in food production. It emphasizes recycling techniques and low external input and high output strategies. It is based on enhancing soil fertility and diversity at all levels and makes soils less susceptible to erosion. In this publication, organic farming and food systems are evaluated in the context of climate change scenarios. As simple answers cannot be given to such a complex and global problem, it is equally important to highlight recommendations for future development and research requirements in organic agriculture.

Organic farming links productivity with ecology and creates livelihoods in rural areas: it is a surprisingly multifaceted option. The current change in global climate is a phenomenon that is largely due to the burning of fossil energy (coal, oil, natural gas) and to the mineralisation of organic matter as a result of land use. These processes have been caused by mankind's exploitation of fossil resources, clearing of natural vegetation and use of these soils for arable cropping.

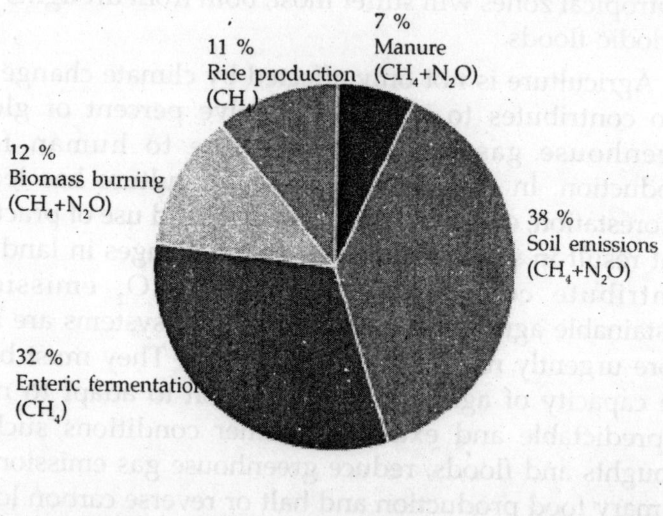


*Figure 1: Main sources of greenhouse gas emissions in the agricultural sector in 2005*

These activities have primarily led to a measurable increase in the carbon dioxide ($CO_2$) content of the atmosphere, an increase which results in global warming, as $CO_2$ hinders the reflection of sunlight back into space, and thus more of it is trapped in the Earth's atmosphere. Molecules of methane ($CH_4$) and nitrous oxide ($N_2O$) have a similar, but far greater effect: the global warming potential of methane is twenty times that of $CO_2$, while that of nitrous dioxide is as much as 300 times greater.

IPCC has published greenhouse gas emissions classified by different sectors as shown in figure 1. When calculating the climate impact of a certain production type it is always a question, where to put the cut-off points of a particular system. For instance, agricultural emissions as shown in figure 1 do not comprise emissions from fertiliser production, which are counted under 'industry'. This needs to be taken into account when comparing farming systems. When considering the total food chain from the farm to the consumer, emissions from all the other sectors need to be included. Thus, the greenhouse gas emissions from all sectors related to agriculture may potentially sum up to 25-30% of all GHG emissions.

## GREENHOUSE GASES EMISSION IN AGRICULTURAL SECTOR

According to the Intergovernmental Panel on Climate Change (IPPC), the annual amount of greenhouse gases emitted by the agricultural sector is estimated at between 5.1 and 6.1 gigatonnes $CO_2$ equivalents in 2005. This represents approximately 10–12% of total greenhouse gas emissions. Of these emissions, methane accounts for 3.3 Gt equivalents and nitrous oxide for 2.8 Gt $CO_2$ equivalents annually, while net emissions of $CO_2$, at only 0.04 Gt $CO_2$ equivalents per year, are small. Agriculture is the main emitter of nitrous oxides and methane according to current practice and knowledge.

Emissions of nitrous oxide originate mainly from:

— high soluble nitrogen levels in the soil from synthetic and organic nitrogen sources (fertilisers).

— animal housing and manure management.

The main sources of methane emissions are:

- — enteric fermentation by ruminants (e.g. cows, sheep, goats).
- — anaerobic turnover in rice paddies.
- — manure handling.
- — compaction of soils resulting from the use of heavy machinery.
- — biomass burning, e.g. from slash-and-burn agriculture, emits both methane and nitrous oxide.

Vegetation – together with the soil ecosystem as the place for decomposition – generates large fluxes of carbon dioxide ($CO_2$) both to and from the atmosphere. According to the Intergovernmental Panel on Climate Change (IPPC), this flux is nearly balanced in agriculture with a net emission of 0.04 Gt $CO_2$ equivalents per year, and represents less than 1% of global anthropogenic $CO_2$ emissions.

Substantial emissions of $CO_2$ from soils, however, originate from land-use changes such as deforestation (not counted under the agricultural sector by IPCC). On the other hand, reforestation and afforestation are considered sinks for $CO_2$. By sequestering carbon dioxide in the soil, agriculture may contribute to the carbon cycle in a positive way. Whether the balance is positive or negative depends to a large extent on the cutoff points for the comparison (for instance if the fertiliser industry is included or not) as well as the management and farming practices applied. Agriculture has the potential to be a considerable $CO_2$ sink, if good farming practices, like organic farming, are employed.

Predictions concerning the future global trends for greenhouse gas emissions from agriculture largely depend on physical and economic parameters that have a strong influence on total emissions. These parameters include: cost of fuel, economic development, evolution of livestock numbers, increase in productivity, new technology, availability of water, deforestation, and consumer attitudes and diet.

According to current projections, total greenhouse gas emissions from agriculture are expected to reach 8.3 Gt $CO_2$ equivalents per year in 2030, compared to the current level of approximately 6 Gt $CO_2$ equivalents annually.

## RISKS OF THE UNPREDICTABLE WEATHER

Current scientific models predict substantial environmental changes caused by increased emission of greenhouse gases. These changes will affect agriculture both in positive and negative ways. The forecast increase in global temperature of between 1.4°C and 5.8°C will result in alterations in precipitation patterns.

Extreme weather events (droughts, floods) are expected to occur more frequently. Seasonal variations in weather events may pose risks to traditional methods of crop production either due to water constraints or surplus of water and erosion. Soil stability will become crucial in order to store water in the soil profile, to resist severe weather events and minimise soil losses.

Vulnerable regions such as tropical and subtropical areas and high mountain regions are expected to suffer most from climate change.

## MEASURES TO MITIGATING THE GLOBAL WARMING IMPACT

The Intergovernmental Panel on Climate Change IPPC has suggested a range of measures for mitigating greenhouse gas emissions from agricultural ecosystems. Ac-cording to Barker et al., sink enhancement (carbon sequestration) will contribute most to mitigation in this context. Prominent mitigation options in agriculture according to IPCC include improved cropland management (including nutrient management, tillage/residue management and water management), improved grazing land management (e.g. grazing intensity, nutrient management) and the restoration of degraded soils.

Agriculture can help to mitigate climate change by a) reducing emissions of greenhouse gases (GHGs) and b) by sequestering $CO_2$ from the atmosphere in the soil. The potential of organic agriculture for both effects is high, as data gained from modelling both long-term field trials and pilot farms show.

### Reduction of Greenhouse Gas Emissions

#### *Organic farming has lower global warming potential*

The global warming potential (GWP) of agricultural activities can be defined as greenhouse gas (GHG) emissions in $CO_2$ equivalents per unit land area or per unit product. The global warming potential of organic farming systems is considerably smaller than that of conventional or integrated systems when calculated per land area. This difference declines, however, when calculated per product unit, as conventional yields are higher than organic yields in temperate climates. Under dry conditions or water constraints, organic agriculture may outperform conventional agriculture, both per crop area and per harvested crop unit.

Organic farms do not only produce cash crops, but they use arable fields for temporary grass-clover (ley-farming) and fodder production for livestock. A part of the grass-clover yields and nutrients is used to fertilise arable crops and soils. Depending on the methodology and the calculations used, the results can differ substantially:

— In the two long term comparison experiments with arable rotations in Switzerland, the DOK trial and the Burgrain experiment, the global warming potential of all crops was reduced by 18% in the organic plots. A reduction of warming potential has also been found in

Dutch dairy farms and in some vegetable crops.

— In contrast, no difference or higher emissions for organic crops were modeled for the experimental farm in Scheyern (Bavaria, Germany), for 18 Bavarian commercial crop farms and for most of the vegetable crops from model farms in the Netherlands.

When losses and gains of soil carbon stocks (mineralisation or sequestration) are considered in the calculations, the global warming potential is considerably reduced for organic agriculture as shown in recent studies:

— Scheyern experimental farm: decrease of 80%.

— Bavarian survey of 18 commercial farms: 26%.

— Station experiment in Michigan: 64%.

***Role of nitrogen and $N_2O$***

*Synthetic nitrogen fertilisers as a major contributor to global warming*

The global warming potential of conventional agriculture is strongly affected by the use of synthetic nitrogen fertilisers and by high nitrogen concentrations in soils. Global nitrogen fertiliser consumption (produced by fossil energy) in 2005 was 90.86 million tonnes. It takes approximately 90 million tonnes of fossil fuel (diesel equivalents) to produce this nitrogen fertilizer. This is about 1% of global fossil energy consumption. In the UK, a 100-hectare stockless arable farm consumes on average 17,000 litres of fossil fuel annually through fertiliser inputs.

*Self-sufficient in nitrogen*

Organic agriculture, in contrast, is self-sufficient in nitrogen. Mixed organic farms practice highly efficient recycling of manures from livestock and of crop residues by composting. Leguminous crops deliver additional nitrogen in sufficient quantities (on stockless organic farms this is the main source). Badgley et al. calculated the potential nitrogen production by leguminous plants via intercropping and off-season cropping to be 154 million tonnes, a potential which exceeds the nitrogen production from fossil fuel by far and which is not fully exploited by conventional farming techniques.

*Reduced emissions of nitrous oxide*

Emissions of nitrous oxide are directly linked to the concentration of easily available mineral nitrogen in soils. High emission rates are detected directly after fertilisation and are highly variable. Denitrification is additionally enhanced in compacted soils. According to IPCC, 1.6% of nitrogen fertiliser applied is emitted as nitrous oxide. In organic agriculture, the ban of mineral nitrogen and the reduced livestock units per hectare considerably reduce the concentration of easily available mineral nitrogen in soils and thus $N_2O$ emissions.

Furthermore, these factors add to lower emissions of nitrous oxide:

- Diversified crop rotations with green manure improve soil structure and diminish emissions of nitrous oxide.
- Soils managed organically are more aerated and have significantly lower mobile nitrogen concentrations. Both factors reduce emissions of nitrous oxides.

In the study by Petersen et al, lower emission rates for organic compared to conventional farming were found for five European countries. In a long-term study in southern Germany, Flessa et al. also found reduced nitrous oxide emission rates in the organic farm, although yield-related emissions were not reduced.

*Integration of livestock and crop production*

The on-farm use of farmyard manure – a practice increasingly abandoned in conventional production –

needs to be reconsidered in the light of climate change. While conventional stockless arable farms use synthetic nitrogen fertilisers, manure and slurry from dairy, beef, or from non-ruminant farms have become an environmental problem. In these livestock operations, nutrients are available in excess and overfertilisation occurs. Emissions of $CO_2$, nitrous oxide and methane are likely to be very high and water pollution may occur when manure is treated as waste and not as recycled as a valuable fertiliser in the crops.

Integration of livestock and arable production, the rule on organic farms, can thus reduce the global warming potential of food production. This fact is not calculated correctly in most global warming potential models, however, as livestock production is generally considered separately from crop systems.

*Nitrogen efficiency as a key factor for the reduction of greenhouse gases*

Greenhouse gas emissions at farm level may be related either to the farm's nitrogen surplus or to the farm's nitrogen efficiency, as demonstrated by a scientific model of greenhouse gas emissions from European conventional and organic dairy farms. Farm nitrogen surplus can therefore be a good proxy for greenhouse gas emissions per unit of land. Since organic crop systems are limited by the availability of nitrogen, they aim to balance their nitrogen inputs and outputs and their nitrogen efficiency. Their greenhouse gas emissions are thus lower than those of conventional farming systems.

***Methane emissions***

Methane accounts for about 14% of the greenhouse gas emissions. Twothirds of this are of anthropogenic origin and mainly from agriculture. Methane emissions stem to a large extent from enteric fermentation and manure management and in consequence are directly proportional to livestock numbers. Avoidance of methane emissions of anthropogenic origin and especially of agricultural origin is of particular importance for mitigation. Organic agriculture has an important, though not always superior, impact on reduction as livestock numbers are limited in organic farms.

*Longevity of animals on organic farms contributes to reduction of methane emissions*

Organic cattle husbandry contributes positively to reducing methane emissions by aiming towards animal longevity. The ratio between the unproductive phase of young cattle and the productive phase of dairy cows is favourable in organic systems because, calculated on the basis of the total lifespan of organic dairy cows, less methane is emitted. On the other hand, lower milk yields of organic cows caused by a higher proportion of roughage in the diet, might increase methane emissions per yield unit. A model calculation of the best yield-methane emission rate at different diets (roughage versus concen-trates) is missing. The slightly reduced yields of organic farms might be nearer the optimum than conventional dairy production.

*Composting and biogas production as measures for mitigating climate change*

Composting and biogas production are often suggested as measures for mitigating climate change. In the context of climate change, the benefits of aerobic fermentation of manure by means of composting are ambiguous, as a shift from anaerobic to aerobic storage of manure can reduce methane emissions, but will increase emissions of nitrous oxide by a factor of 10. A very promising option, however, is controlled anaerobic digestion of manure and waste combined with biogas production. While this option is not restricted to organic production methods, organic agriculture has been at the forefront of biogas production systems for decades. Attention must be paid however to the economic viability of biogas production systems.

*Methane emissions from organic rice production and from ruminants: Improved techniques needed*

Methane emissions from organic rice production and ruminant production do not differ substantially from those of conventional production. Better rice production practices in organic and conventional agriculture, such as avoiding continuous flooding or choosing low methaneemitting varieties could enhance reduction of methane emissions. The multi-target approach of organic farmers and the fact that they are often more highly skilled could enhance implementation of improved production techniques.

## Organic Farming Sequesters $CO_2$ in the Soil

*Soil erosion results in loss of soil carbon*

Arable cropland and permanent pastures lose soil carbon through mineralisation, erosion (water and wind-driven) and overgrazing. Global arable land loss is estimated at 12 million hectares per year, which is 0.8% of the global cropland area (1513 million hectares). This rapid loss is confirmed by experimental data from Bellamy et al. in England and Wales. Between 1978 and 2003, they found carbon losses in 92% of 6000 soil samples. Annual $CO_2$ emissions from intensively cropped soils were equivalent to 8% of national industrial $CO_2$ emissions.

If agricultural practices remain unchanged, the loss of organic carbon in typical arable soils will continue and eventually reach a new steady state at a low level. The application of improved agricultural techniques (e.g. organic farming, conservation tillage, agroforestry), however, stops soil erosion and converts carbon losses into gains. Consequently, considerable amounts of $CO_2$ may be removed from the atmosphere.

*Carbon gains*

Organic land management may help to stop soil erosion and convert carbon losses into gains, particularly due to:

— the use of green and animal manure.
— soil fertility-conserving crop rotations with intercropping and cover cropping.
— composting techniques.

*Higher soil organic matter content in organic farming*

Farm comparison and long-term field trials show that organically managed soils have significantly higher organic matter content. It is estimated that, under Northern European conditions, conversion from conventional to organic farming would result in an increase of soil organic matter of 100 to 400 kg per hectare annually during the first 50 years. After 100 years, a steady state, i.e. a stable level of soil organic matter, would be reached.

*Carbon sequestration rates on organic farms*

Under 'real' conditions in long-term experiments, or in farm-level comparisons, carbon sequestration rates vary considerably.

*Organic versus conventional soil conservation strategies: No-tillage and minimum-tillage cropping*

In the past decade, agricultural techniques have been developed to maintain soil fertility and soil quality. By reducing the intensity of tillage, soil conservation can be improved, and water and wind erosion can be considerably reduced.

— Robertson et al. compared the greenhouse warming potentials (GWP: including carbon sequestration, agronomic inputs and trace gas emissions) of conventionally tilled, no till and organic farming systems in the Mid Western US and found none of these agricultural systems to be climate neutral. Whereas no-till reduced the GWP of conventional tillage by 88%, organic production with legume cover was only 64% lower than conventional tillage.

— In a nine-year system comparison experiment in Beltsville, it was shown that the organic farming approach provided excellent soil fertility building and was superior to conventional no-tillage techniques, despite the use of a plough.

No-tillage cropping is mainly practised on stockless farms, which leads to highly specialised farms – either crops or animals – and excess manure on the animal farms becomes an environmental problem. Nitrate excess in the soil triggers emissions of nitrous oxide, as well as nitrate leaching and phosphorus run-off. The organic approach involving local recycling and nutrient use in a mixed-farm approach offers many ecological benefits.

In very fragile soils, it is nonetheless recommended to use minimum-tillage techniques in organic farming as well. Several research projects in different parts of the world are working on such systems. For instance, in Switzerland a long-term trial was recently started that analyses the effect of reduced tillage on crop yields and weed infestation. Similar research projects are running at Bonn University in Germany. Technically, there is no inherent incompatibility between organic and minimum-tillage cropping.

*Agroforestry, Permaculture, Polyculture*

Agroforestry is a management system that integrates perennial and annual crops in a twocanopy or multi-canopy production system. This guarantees better exploitation of light, water and soil nutrients and protects soil more effectively from erosion and leaching. It leads to a more diversified and sustainable production system than many treeless alternatives and provides increased social, economic and environmental benefits for land users. In the humid tropics, agroforestry is seen as viable alternative to slash-and-burn agriculture.

The $CO_2$ sequestration potential of agroforestry in the short and medium term is mainly above ground. The additional carbon in the standing vegetation may increase by 50 tonnes per hectare; in the soil by an additional 7 tonnes per hectare. These increases were measured 20 to 25 years after recultivation of previously cleared forests..

Even though agroforestry is neither restricted nor exclusive to organic agriculture, organic principles suit it very well. Consequently, organic agriculture could play a role in the development of agroforestry systems and combining these two systems is a potential solution for reducing greenhouse gases, sequestering carbon dioxide and increasing the productivity of agro-ecosystems.

### Mitigation Potential of Organic Agriculture beyond Purely Agricultural Practices

#### *Changing consumer behaviour and diet*

The greatest potential for reducing greenhouse gas emissions from agriculture would be to change consumer behaviour. Production of meat requires inputs that are seven times as high as the inputs needed to produce the same quantity of non-meat calories. Organic agriculture aims at precisely this goal: consumption of less-processed products and increased consumption of products like cereals, potatoes, pulses and oils.

Greenhouse gas emissions are highest in beef production ($CO_2$ equivalents per kg meat are higher than 10,000 g), followed by pork, poultry and egg production

(2,000 to 3,000 g $CO_2$ equivalents per kg) and milk (approximately 1000 g $CO_2$ equivalents per kg). Emissions from production of plant foods are generally below 500 g $CO_2$ equivalents per kg.

Although ruminants (cattle, sheep, goats) are major methane emitters, they are crucial to global food security as they tap into an area of 3,432 million hectares worldwide that is not suitable for crop production. These animals carry bacteria in their rumen that make plant material digestible, that other animals are hardly able to use, but unfortunately this fermentative process is also emitting methane.

### *Stopping deforestation*

In organic farming, preparation of the land by burning vegetation is restricted to a minimum. Organic farming thus contributes to halting deforestation with its highly negative impact on climate change. Often, the opposite argument has been made, as organic agriculture usually needs more land to produce the same amount of food as by conventional farming. This might be compensated by the potential of organic agriculture for aiding reclamation and making use of degraded land due to its favorable effects on soil fertility and soil organic matter. In addition careful land use and management as in organic farming enhances environmental security and will help to stop losses of fertile arable land not only by erosion.

## POTENTIAL TO ADAPT CLIMATE CHANGE

Agricultural production in most parts of the world will face less predictable weather conditions than mankind experienced during the intensification of agriculture over the last century. Weather extremes will become predominant. Resilience and adaptiveness are new requirements gaining importance for innovation in agriculture.

The Intergovernmental Panel on Climate Change (IPPC) defines adaptation to climate change as 'adjustment in natural or human systems in response to actual or expected climatic stimuli or their effects, which moderates harm or exploits beneficial opportunities'.

### Traditional Skills and Knowledge

Traditional skills and knowledge have been neglected in intensive agriculture, although they are now being partially recaptured by integrated pest management. Organic agriculture, on the other hand, has always been based on practical farming skills, observation, personal experience and intuition. Knowledge and experience replaces or reduces reliance on inputs. This knowledge is important for manipulating complex agro-ecosystems, for breeding locally adjusted seeds and livestock, and for producing on-farm fertilisers (compost, manure, green manure) and inexpensive nature-derived pesticides. Such knowledge has also been described as a 'reservoir of adaptations'.

### Organically Managed Soils

Farming practices such as organic agriculture that preserve soil fertility and maintain or even increase organic matter in soils are in a good position to maintain productivity in the event of drought, irregular rainfall events with floods, and rising temperatures. Soils under organic management retain significantly more rainwater thanks to the 'sponge properties' of organic matter.

- These 'sponge properties' were described for heavy loamy soils in a temperate climate in Switzerland where soil structure stability was 20–40% higher in organically managed soils than in conventional soils.
- The amount of water percolating through the top 36 cm was 15–20% greater in the organic systems of the Rodale farming systems trial compared to conventional

systems. The organic soils held 816,000 litres per ha in the upper 15 cm of soil. This water res-ervoir was likely the reason for higher yields of corn and soybean in dry years.

— It was found that water capture in organic plots was twice as high as in conventional plots during torrential rains. This significantly reduced the risk of floods, an effect that could be very important if organic agriculture were practised on much larger areas.

### Enhancing Productivity of Degraded Soils

Experience with degraded soils of the arid tropics have shown that agricultural productivity can be enhanced using soil fertility building techniques. In the Tigray province of Ethiopia, one of the most degraded parts of the country, agricultural productivity was enhanced by soil fertility techniques such as compost application and introduction of leguminous plants into the crop sequence. By restoring soil fertility, yields were increased to a much greater extent both at farm and regional level than by using bought mineral fertilisers. This large-scale experiment underlines the importance of organic matter and soil fertility for ensuring productivity in dry regions and partly explains the surprisingly high yields from organic crops found by Badgley et al.

### Diversity Enhances Farm Resilience

An additional strength of organic farming systems is their diversity – including the diversity of crops, fields, rotations, landscapes and farm activities (mix of various farm enterprises). The high level of diversity of organic farms provides many ecological services that significantly enhance farm resilience. Positive effects of enhanced biodiversity on pest prevention have been shown by several authors. Similar effects of diversified agro-ecosystems on diseases and better utilisation of soil nutrients and water are likely to occur.

## WEAKNESSES OF ORGANIC AGRICULTURE IN THE CLIMATE CHANGE

### Organic Farming is Less Productive

#### *Agronomically difficult crops as a challenge*

One major criticism of organic agriculture is that productivity is lower compared to intensive conventional agriculture. Under geoclimatic conditions that allow for a very high yield, in the case of some crops the relative advantage of organic agriculture in terms of energy consumption per land area compared to conventional production may switch to the contrary when calculated on the basis of crop or livestock yield. This is particularly true in the case of highly demanding crops such as potatoes, grapes fruits and horticultural crops especially from greenhouse production. Pest, disease and weed management problems relating to these crops have not yet been resolved satisfactorily. Consequently, the yield for these crops is too low and the energy input, even though relatively low on a land area basis, becomes relatively high when applied to a crop unit. With more research into organic agriculture, however, progress may be expected in this field.

#### *Better technology transfer could improve organic yields*

The productivity of organic agriculture is often underestimated by many scientists and policymakers. Organic agriculture represents a very productive food supply system that relies on recycling strategies. Badgley et al. modelled the yields stated in 293 on-farm and onstation publications and concluded that, compared to high-input agriculture in developed countries, the average yields in organic crop and livestock production are 92% of those in conventional agriculture.

As scientific progress in conventional agriculture has accelerated in recent years, if we look only at the most recent data and at selected countries in temperate climate zones, yield differences as obtained from official databases are generally higher. In the case of wheat, for example, organic yields vary between 58% and 98% of those of the conventional pair. This rather wide range of yields measured by farm surveys indicates that yields on organic farms might be improved considerably, e.g. by better technology transfer from research to practice.

*Organic farming performs better under water constraints*

Interestingly, yields from organic agriculture under conditions where water is limited during the growing period, and under subsistence farming, are equal or significantly higher than those from conventional agriculture. A comparison of 133 studies from developing countries concluded that organic plant and livestock yields were 80% higher than their conventional counterparts (for crops only the increase in yield was by 74%).

In temperate climate zones higher yields for maize and soya were registered in organically managed fields in dry seasons. The available data indicates that the technique inherent to organic farming of investing in soil fertility by means of green manure, leguminous intercropping, composting and recycling of livestock manure could contribute considerably to global food productivity. Further improvements in manure storage and application techniques are also required, however, in organic systems in order to reduce nitrous oxide and methane emissions.

**High Dependency on Nutrients Derived from Livestock**

Some critics are concerned about the dependency of organic cropping upon nutrients deriving from livestock. This criticism, however, underestimates manure as a valuable and potentially useful resource. Moreover, this is not a relevant weakness since the numbers of animals kept in agriculture depends mainly on consumer demand.

In order to reduce greenhouse gas emissions, efficient and direct recycling of manure and slurry is the best option, since it avoids long-distance transport and consumption of energy for synthetic fertiliser production. The combination of crop and livestock production is currently the most efficient way of bringing organic 'waste' from livestock production back into the carbon stock of the soils and use it as a locally available resource for crop fertilisation and enhancing soil quality.

When integrated into arable farming systems, ruminants exploit leguminous crops and intercrops that are needed to produce nitrogen, provide soil cover and capture soluble nutrients, building up soil fertility and soil structure.

From the point of view of integrating ecology and sustainable resource use, the combination of organic cropping with livestock production is undoubtedly a strong point.

### Need More Funding for Research on Organic Farming

As 99 % of the world's public and private research funds have focussed on optimising conventional and integrated food and farming systems during the last decades, major progress and solutions can be expected as a result of agro-ecological and animal welfare research activities.

## CLIMATE CREDIT FOR ORGANIC FARMING

Agriculture, both organic and conventional, has the potential to make a cost-effective contribution to mitigation. The Intergovernmental Panel on Climate Change (IPCC) estimates that agricultural greenhouse gas mitigation options are costcompetitive with non-agricultural options in achieving long-term climate objectives.

### Special Benefits of Organic Agriculture

Within agriculture, organic agriculture holds an especially favourable position, since it realises mitigation and sequestration of $CO_2$ in an efficient way. Compared to other agricultural systems, organic farming is a well-defined system that is already based on certification and that could easily be extended to meet the standards of CDM.

Organic production has great mitigation and adaptation potential, particularly with regard to soil organic matter fixation, soil fertility and water-holding capacity, increasing yields in areas with medium to low-input agriculture and in agroforestry, and by enhancing farmers' adaptive capacity. Paying farmers for carbon sequestration may be considered a win-win-win situation as a) $CO_2$ is removed from the atmosphere (mitigation), b) higher organic matter levels in soil enhance their resilience (adaptation) and c) improved soil organic matter levels lead to better crop yield (production).

Improvement is needed, however, with regard to yields and methane emissions. Organic agriculture, with its holistic multi-target approach, offers further relevant advantages with regard to lifestyle changes, for example, primarily in developed countries. Its numerous cobenefits could greatly assist the development of rural societies in southern countries.

It would benefit the cause if agriculture, and in particular organic agriculture, could be included as a high-benefit / low-cost $CO_2$ reduction system in the next climate agreement, negotiations on which, referred to as the 'post 2012 negotiations', are about to begin.

### Clean Development Mechanism

Agriculture, however, and organic agriculture in particular, are barely able to benefit from climate credits, of which the Clean Development Mechanism (CDM) is the most popular. Even though agriculture could apply for CDM credits for reduction of methane through biogas installations for example, the single most important measure in terms of greenhouse gas reduction, sequestration of $CO_2$ in soils, is excluded from the CDM mechanism.

Likewise, the 'gold standard' launched by the World Wide Fund for Nature (WWF) formally excludes agricultural $CO_2$ sequestration sinks from eligibility for carbon credits. For both organisations, the reason is that agricultural $CO_2$ sinks are considered to be temporary in nature only. Once $CO_2$ has been sequestered, a change in land use poses the risk that a substantial amount of CO2 will be released back into the atmosphere. The logic of this perception leads to a situation where only afforestation or reforestation projects are supported. Organic mixed farms, especially when good compost is used, tend to build up organic matter with a higher proportion of stabile fractions. Therefore, quitting organic farming might not lead to a faster release of $CO_2$ than cutting a tree plantation.

## REFERENCES

Flessa, H., Ruser, R., Dörsch, P., Kamp, T., Jimenez, M.A., Munch, J.C., Beese, F., "Integrated evaluation of greenhouse gas emissions ($CO_2$, $CH_4$, $N_2O$) from two farming systems in southern Germany", *Agriculture, Ecosystems and Environment*, 2002.

Hole, D.G., A.J. Perkins, J.D. Wilson, I.H. Alexander, P.V. Grice and A.D. Evans, "Does organic farming benefit biodiversity?", *Biological Conservation*, 2005.

Holland, J.M., "The environmental consequences of adopting conservation tillage in Europe: reviewing the evidence", *Agriculture, Ecosystems and Environment*, 2004.

Kotschi, J., Müller-Sämann, K., "The Role of Organic Agriculture in Mitigating Climate Change", *International Federation of Organic Agriculture Movements (IFOAM)*, Bonn, 2004.

Reganold, J., Elliott, L. and Unger, Y., "Long-term effects of organic and conventional farming on soil erosion", *Nature*, 1987.

# 11

# Organic Farming and Food Safety

By producing food, farmers can improve or degrade the environment they depend on. Both favorable and stable environmental conditions and agricultural systems resistant or resilient to environmental changes are crucial for the stability of food production. Stability of food supply is also associated with other dimensions of food security, such as access to food and food utilisation, as well as economic conditions of food stability. Organic agriculture here includes both those production systems organised around distinct organic and biodynamic certification schemes that are "certified organic agriculture" and those that, by virtue of their production systems, follow organic principles but are "non-certified organic".

Organic agriculture is a system that uses less intensive practices. It is, therefore, expected to deliver far more ecological goods and services than conventional agriculture, even in its modified forms such as integrated farming or minimum tillage. Furthermore, organic agriculture efficiently reduces environmental risks by not using some potentially damaging technologies of intensive agriculture such as pesticides, herbicides, synthetic nitrogen fertilisers, GMO crops or veterinary antibiotics and anthelminthics.

In broad terms, organic agriculture can be described as increased diversification on both farm and regional scale.

It is a system that leads farmers and rural communities to economic independence from expensive agricultural inputs through their use of participatory seed-breeding systems; natural bio-control agents; soil fertility management through recycling, nitrogen fixation and green manure; and habitat management as a prevention strategy against pests, diseases and weeds.

Currently, 0.61 percent of the world's reported agriculture land is certified organic agriculture. There are no recent data on the extent of non-certified organic agriculture. However, it is widely known that a large part of global food production systems is non-certified organic agriculture, often at subsistence level.

### ENVIRONMENTAL INFLUENCE ON FOOD SUPPLY STABILITY

The key environmental influences on the performance of agricultural systems are soil and water resources, biodiversity and climate. Agriculture is both perpetrator of changes – in most cases environmental deterioration – and victim. Among all concepts of sustainable land use practices, organic agriculture is the most consequential approach, in terms of ecological sustainability.

With respect to the quality of sustainability of organic agriculture, two observations can be made.

— Sustainability from organic systems is based on standards that advocate restrictions or bans on the input side. However, metrically measurable standards on the impact side (e.g. minimum percentage of high quality ecological areas such as hedge rows, field margins and species rich wildflower strips, or minimum of crops in a rotation) would improve the sustainability of organic agriculture even more. As they are very region specific, such standards focusing on positive environmental impacts can only be introduced in regional or national standards as has been done in Switzerland.

— Organic agriculture is a multi-targeted approach to sustainability. Unlike conventional farming, it does not focus on individual high impact measures. For example, the "no till" technique used in conventional farming efficiently prevents soils erosion but needs more herbicides and soluble nitrogen fertilisers and often promotes soil borne diseases leading to need for additional fungicidal treatments

## Stability and Fertility of Soils

In order to meet the world's current and future food demand, food supply stability will be needed to maintain agricultural productivity at least at the current level without negative effects on soils or the environment. In addition, the expanding human population will require more secure and improved yields and additional acreage. Yield increases resulting from the so-called "Green Revolution" have slowed and are currently linked to soil degradation, which is considered a threat to food supply stability. Pimentel, et al. calculated a loss of nearly a third of the world's arable land to erosion within the last 40 years with an on-going loss of more than 10 million ha per year. Bellamy, et al. found massive losses of carbon in soils across England between 1978 and 2003. Their estimates ranged from 0.5 to 2 g soil carbon per kg soil per year with all but 8 percent of the investigated cropland affected by erosion – a factor the authors identified as the main reason for losses in soil carbon and therefore in soil fertility. This highlights that current land use practices are not sustainable.

Reganold, et al., in comparing soils from organic and conventional farms in Washington, USA, found organic fields had topsoils 16 cm deeper and a higher organic matter content which resulted in soils less prone to erosion. A long-term Swiss field experiment on loess soil that began in 1978 found the aggregate and percolation stability of both bio-dynamic and organic plots were significantly higher (10 to 60 percent) than conventionally farmed plots. This also affected the water retention potential of these soils in a positive way and reduced their susceptibility to erosion. Soil aggregate stability was strongly correlated to earthworm and microbial biomass, important indicators of soil fertility.

The long-term application of organic manure positively influenced soil fertility at the biological, chemical and physical level, whereas the repeated spraying of pesticides appeared to have negative effects. Compared to stockless conventional farming (mineral fertilisers, herbicides and pesticides), repeated measurements of aggregate stability in plots with livestock-based integrated production (mineral and organic fertilisers, herbicides and pesticides) found 29.4 percent higher values while in organic and bio-dynamic plots organic fertilisers only), it was 70 percent higher. The Swiss long-term study underlines the importance of using manure, by means of organic agriculture, as a good practice for soil quality preservation.

Similar results were obtained under on-farm conditions in the Netherlands in a polder soil, which is considered prime agricultural land for Dutch arable crop production. The percentage of water stable macro-aggregates on organically farmed sites was 72 percent higher compared to conventional. The higher physical stability was linked to significantly increased soil organic matter content and to a larger volume percentage of worm-worked soil. The investigation was done on farms that had been under organic and conventional management for 70 years. A study of cotton production under organic conditions in India found yield levels similar to a modern cultivation technique, but soil quality, as indicated by soil organic matter, water stable aggregates and mean weight diameter, showed advantages for the organic system.

The Rodale farming systems trial, that began in 1981 in Pennsylvania, USA, compared manure and legume-based organic agriculture systems to a conventional system based on mineral fertilisers. It found the organic and conventional systems had similar soybean and maize yields whereas the organic system showed an impressive increase in soils carbon of 574 kg per ha in the legume-based and 981 kg ha-1 in the manure-based system.

Soil macrofauna, such as worms, ants and termites that are actively involved in the build-up of soil structure, have positive effects on water infiltration, drainage, water-holding capacity and soil aeration. These and other beneficial soil biota help process nutrients from residues for plant uptake while also creating stable organic matter. Several field studies found earthworms and other soil fauna such as carabides, spiders and staphylinides were more abundant in organic fields than conventional ones. A more diverse food and micro-landscape (weeds and soil structure) and the absence of potentially toxic pesticides are the main factors favouring macrofauna in organic systems.

The predominant practice in organic agriculture is the on-farm flux of manure from livestock production to cropland. However, even stockless organic farms use leguminous plants for nitrogen fixation and green manure for building up soil fertility. Catch crops for soil nutrients, especially nitrogen, integrated in different positions of the rotations, help manage nitrogen intelligently, especially in temperate climatic zones where nitrogen losses are likely to occur.

Soil organic matter, a key to soil fertility throughout agriculture, is understood to contribute significantly to soil quality and health. Soil organic matter drives and enhances numerous chemical and biological processes and physical soil properties. Higher levels of soil organic matter have been observed regularly in organic agriculture soils compared to conventional. In one study, soil organic matter was higher in organically managed soil than in conventional soil despite relatively similar totals of organic carbon.

Marriott and Wander analysed soil samples from nine farming system trials that were started in the USA between 1981 and 2000. The soil organic carbon concentrations were 14 percent higher in organic systems than in conventional ones. The labile fraction of the soil organic matter – a source of mineralisable C and N with important implications for plant nutrition – showed 30 to 40 percent higher values in organic soils. Enhancing soil organic matter improves soil fertility and, in the long-term, has the potential to increase crop yields.

Parallel to the changes of organic matter in soils of organically managed systems, soil microbial biomass and the physiological functions of soils are enhanced by organic agriculture. Important soil enzyme activities such as dehydrogenase, protease and phosphatase were higher in organic field plots leading, for example, to a faster phosphorus flux through the microbial biomass contributing to the plants' phosphorus supply. Field trial soils that received only mineral fertilisers in a conventional system showed a 52 percent higher basal respiration per unit of microbial biomass than organic soils, suggesting that micro-organisms in organic soils use organic substances more efficiently for growth than for maintenance.

In short, organic practices contribute considerably to increasing soil stability and resilience, the latter reflects the time needed to recover from disturbances, an important factor in food supply stability. The productivity of organic compared to conventional farming depends strongly on soil and climate conditions as well as on choice of crops being compared. Under less favorable soil conditions, organically managed crop yields equal those from conventional agriculture, as was shown for maize and soybeans in the USA by Pimentel, et al. In organic cotton production in India, Eyhorn, et al. also reported no yield differences from organic compared to conventional farms.

In a fertile soil in a temperate climate in Switzerland, total yields of the organic 7-year crop rotation were 20 percent lower than the conventional one, but had a significantly better output/input ratio for nutrients, energy and pesticides in the case of organic practice. When soil and climate conditions are extremely favorable, yield losses of organic crop rotations may exceed 30 percent of conventional ones. This may also apply to crops with peculiar plant protection demands and short vegetation periods.

In comparison to other modern approaches to improving crop yields and soil quality in order to prevent soil erosion such as minimum or no tillage systems, organic agriculture advocates a site-adapted multifactorial strategy without the use of synthetic fertilisers or pesticides. Instead, organic agriculture makes use of available production factors that mainly derive from the farm-ecosystem itself. The combination of techniques – organic agriculture and minimum or reduced tillage – is a challenging subject for future research in this field. Long-term field experiments addressing this area are underway (Berner, et al. submitted) and need to be performed under different site and climate conditions.

### Organic Agriculture and Water Efficiency

Organic agriculture practices have the potential to use less water and to use water more efficiently due to vegetation and soil management. Soil and water use are inextricably linked in agricultural production. During the twentieth century, humans increased the diversion of river water by six-fold. By 2025, as many as 60 percent of all humans may face a scarcity of water. Currently, 70 percent of human water use is directed towards agriculture. As water becomes increasingly scarce in certain regions of the world, it will be important to increase water efficiency in irrigation and rain-fed agriculture. Water use efficiency is a calculation determined by the amount of crop yielded from an amount of water. Pretty, et al. outlined several ways to improve water use efficiency in organic agriculture, including reducing evaporation through minimum tillage, using more water-efficient varieties and inducing microclimatic changes to reduce crop water requirements.

The soil fertility-building techniques of organic agriculture that lead to higher organic matter contents, better aggregate stability and, biologically, more active soils in turn, increase water retention in soils and improve the water use efficiency. Pimentel et al. reported 28 to 34 percent higher corn yields in the Rodale experiment in Pennsylvania in the organically managed plots in years with drought. During the 2003 summer drought in Europe, similar observations were made with wheat although not scientifically analysed.

Additionally, the presence of soil macrofauna, which are promoted through organic agriculture, can enhance soil water availability. A few studies have shown that organically managed agriculture fields have less water run-off. In the Rodale experiment, water volumes percolating through the organic systems were 15 to 20 percent higher than through the conventional one, indicating an increase in groundwater recharge and reduced run-off. Similar data was obtained in the DOK trial in Switzerland where percolation stability of organic and bio-dynamic soils was 40 percent higher than of conventional ones.

In general, different agricultural techniques used in organic agriculture affect water use efficiency. Tillage, soil residue management, soil temperature, soil nutrients and structure effect water use efficiency. Different degrees of tillage can either increase or decrease water infiltration while also increasing the rate of soil drying. Maintenance of either a cover crop or harvest residue can maintain soil moisture and water use efficiency by increasing water infiltration and cooling the soil surface to a greater degree than tillage that incorporates stubble in the soil. Thus, no-till planting that maintains a cover crop while minimally disturbing the soil surface can increase water use efficiency.

Unfortunately, field experiments comparing the effect of well managed organic systems on water use efficiency with that of good conventional no-till systems do not exist. It can be assumed that organic systems with elements of minimum tillage would be very powerful in terms of water and soil erosion management. In India, although not scientifically analysed observation from biodynamic systems suggest decreased irrigation need by 30 to 50%: this is due to characteristics such as significantly superior soil structure, friability, aeration and drainage, lower bulk density, higher organic matter content, soil respiration (related to soil microbial activity), more earthworms and deeper topsoil layer.

Organic management practices also decrease pollution in water effluent. In addition to maximising water use efficiency in agriculture, it will also be important to decrease the amount of water polluted by agricultural practices for the sake of downstream users and aquatic biodiversity. The prohibition of pesticides and inorganic fertilisers in organic food production is an important step towards removing these pollutants from the aquatic environment. Nitrates are also an agricultural pollutant produced by nitrogen fertilisation, both organic and inorganic.

Organic agriculture, however, has demonstrated reduced nitrate leaching rates, ranking from 40 percent to 64 percent in different soil types in Europe. These effects are less the result of the organic fertilisation technique, as the use of manure and slurry is also prone to nitrate leaching. Reduction in nitrate leaching can only be obtained by carefully designed crop rotations including catch and cover crops. As the opportunity costs of nitrogen on organic farms are 7 to 16 times higher than with mineral fertilisers, these rotational techniques are widely used on organic farms.

### Stability through Biodiversity

Biodiversity is an important driving factor for system stability, and organic agriculture, in particular organic pest and disease management, is based on a most stabilised system. Organic agriculture has been shown to promote more species and have more abundance of organism groups than conventional farming. Stability of agroecosystems can be optimised through the implementation of appropriate soil fertility management and habitat management. Moreover, green manures, application of manure and broad crop rotations within organic farmland lead to greater species diversity and density for insects, plants, soil macrofauna and soil microfauna, although some taxa will not be significantly affected. An over-riding determinant of biodiversity may be habitat diversity, rather than management practices.

Genetic diversity of crops within conventional farms is very low – usually one variety per planting. Organically grown crops can be just as genetically limited but, organic farms have a greater diversity due to mandatory crop rotation. However, one pillar of organic plant protection and animal health is the use of a more diverse range of disease, pest and parasite tolerant or resistant varieties and breeding lines. In most organic crops and in animal husbandry, the use of this genetic diversity is part of the economic strategy to control pests, diseases and parasites.

The potential of genetic diversity on crop level for stabilisation of low-input farming systems and for adaptation to environmental changes is theoretically understood in organic quarters but unfortunately not yet practically used. Cultivar or race choice is still strongly based on immediate market needs; traditional regional or local varieties are niches even in organic agriculture, and in situ concepts of conservation and breeding on farms or in rural communities are still underdeveloped. Kotschi considered agrobiodiversity a fundamental resource for adaptation and therefore crucial for the stability of food supply.

As resistance and robustness to environmental stress are multigenetic characteristics, the in situ conservation and on-farm breeding are likely to be more successful than genetic engineering. There are many very small initiatives by plant and animal breeders in the context of organic farms scattered around the world. These initiatives urgently need political, scientific and economic support.

To show the importance of biodiversity for stability in organic agriculture systems, the strategy of plant protection is exemplary. Organic pest and disease management is based on indirect and direct control measures. Wyss, et al. proposed a conceptual model for arthropod pest management for organic crop production. In this model, indirect, preventative measures are of highest priority. They are to be considered early in the adoption process, followed by more direct and curative measures only when needed. Zehnder, et al. reviewed the preventative and direct measures of this model for organic crops. They concluded that the establishment of an organic production system needs to consider aspects such as landscape complexity to ensure that sufficient semi-natural landscape elements are present to serve as natural enemies that could be attracted by conservation biocontrol methods (e.g. planting hedges, sowing weed strips, installing beetle banks).

Soil quality management (e.g. amendment with compost), tillage practices (e.g. conservation tillage), host plant resistance, crop rotation and intercropping are important additional measures to lower risks of pest and disease outbreaks. Based on just these preventative measures, direct plant protection measures (inundation and inoculation biocontrol, pesticides of vegetal and mineral origin) can be used efficiently by farmers and can therefore be limited. In other words, biodiversity in organic agriculture has a clear function a functional biodiversity that stabilises the agricultural systems.

## ORGANIC AGRICULTURE UNDER CLIMATE CHANGE SCENARIOS

Climate change will dramatically alter international food supplies. Five factors expected with climate change – rise in temperature, changes in precipitation patterns, rise of sea levels, higher incidence of extreme weather events (droughts, storms, floods) and an increase of greenhouse gases (especially carbon dioxide) in the atmosphere – will all influence food production.

Models predict that average global temperatures will increase between 1.8° and 4.0°C at the end of the twenty-first century and will be accompanied by random alterations (both temporal and geographic) in precipitation patterns. These models, which contain a degree of uncertainty, project an increase in agricultural productivity for mid-and high-latitude areas, but a large decrease in low-latitude agricultural productivity. These estimates of net positive and negative effects of climate change on food security take into account the positive effects of increased $CO_2$ on plant growth, reductions in plant water use, and the possibility that new areas for agricultural production will emerge. Negative effects include increased flooding of low-lying areas; increased drought, particularly in tropical countries; and the likelihood of reduced yields in areas affected by decreased precipitation and other changes in climatic conditions.

Land suitable for agricultural land may increase by as much as 40 percent in some developed countries while less developed tropical countries may witness a reduction in agricultural productivity by 5 to 10 percent. As for grazing lands, forage yields are predicted to decrease between 16 and 25 percent. A forthcoming IPPC report on specific global effects of climate change predicts that negative effects will occur disproportionately in equatorial countries compared to temperate ones. The 40 poorest countries located predominantly in tropical Africa and Latin America might lose 10 to 20 percent of their crop production capacity by 2080, mainly due to drought. It is therefore important that resilience of the food supply be maintained and promoted in those places where agriculture is most vulnerable.

Aside from the ability of certain regions to maintain agricultural productivity, irrigation requirements will change with the climate. Fischer, et al. predict that global water withdrawals for irrigation will increase by 20 percent as a result of climate change, in addition to a 25 percent increase that will result from a 45 percent expansion of irrigated cropland in parallel to a 20 percent increase in irrigation efficiency. Around two-thirds of the climatechange-induced increases in water requirements will occur within developing countries.

Combating such changes requires mitigation and adaptation strategies. Above all, all possible efforts are needed to reduce greenhouse gas emissions and to slow climate change. In addition, action is needed to enhance the capacity of agricultural production systems to adapt to irreversible changes. Organic agriculture is a whole food chain approach with unique features that differ from mainstream food and farming production. One can assume that organic agriculture not only enables agroecosystems to better adapt to the effects of climate change but also has high potential to reduce the emissions of agricultural greenhouse gases. It is therefore critical to analyse the major characteristics of organic food and farming systems in the light of their potential for mitigation and adaptation.

## Potential of Organic Agriculture for Migration

Agriculture accounts for 15 percent of the total emission of greenhouse gases (GHG). The whole food chain, including transportation, storage, processing, consumption and preparation, adds up to 20 percent of the total GHG. Of this, agriculture is responsible for emitting one quarter of the carbon dioxide, two-thirds of the methane and 65 to 80 percent of the nitrous oxide.

Afforestation is recognised as a major land use change mitigating climate change. However, crop land management is overestimated for its potential for carbon sequestration especially in Europe. Smith, et al., concluded that "the only trend in agriculture that may be enhancing carbon stocks on croplands at present is organic agriculture, and the magnitude of this effect is highly uncertain."

Organic agriculture offers a wide range of very significant side effects to be seriously considered in the light of reducing GHG emissions in agriculture. Major factors are i) the reduced consumption of fossil fuels for energy, ii) the reduced emissions of carbon dioxide, methane and nitrious dioxide, iii) the considerably reduced vulnerability of soils to erosion accompanied by organic matter respiration, and iv) the increase of carbon stocks which is part of organic management practice and is especially high in already degraded soils.

Energy consumption per land unit has been shown to be 10 to 70 percent lower in all organic crops in five European countries. Due to the lower yields of organic production systems, the energy consumption per product unit is low for wheat, milk, citrus and olive (15 to 54 percent lower in organic systems than in conventional ones) but for potatoes and apples, organic energy use equal or higher than for conventional crops. A study from Iran found organic to be 81 percent more energy efficient than high-input conventional production. For the Rodale long-term experiment in the USA, the energy inputs per ha for the organic systems were 28 to 32 percent less than those of the conventional ones. As the yields of both systems were equal, the relative advantage of organic remained the same per unit produce.

Organic soil fertility management offers a unique possibility to replace fossil-fuel-based nitrogen fertilisers. In the UK, a 100 ha stockless arable farm consumes 17 000 litres of fossil fuel annually through fertiliser inputs. The nitrogen self-reliance of organic systems is a major advantage in times of fossil energy shortage.

Comparative data on the emission of greenhouse gases for organic and conventional systems are rare. Available data show that parallel to the lower energy consumption of organic production, the $CO_2$ emissions are also lower, between 48 and 60 percent. Nemecek, et al. found greenhouse warming potential in organic systems 29 to 32 percent lower on a per ha basis than in a mineral fertiliser system and 35 to 37 percent lower than in the conventional manure-based system. This results from the omission of synthetic fertilisers and pesticides as well as less use of high energy feedstuff.

In most cases, organic systems are also advantageous per unit yield, especially for dairy and meat production and for most arable crops. For technically difficult crops such as potatoes, grape and fruits (with insufficient pest and disease control), as well as for very intensive crops, organic vegetable and arable crop systems have equal or higher $CO_2$ emissions than conventional ones.

As far as $N_2O$ and methane emissions are concerned, no data on organic production is available at the moment. Models used for comparing farms or production units do not differentiate between organic and conventional. Nonetheless, some inherent characteristics of organic systems are likely to lower $N_2O$ and methane emissions. $N_2O$ emission is directly linked to the concentrations of easily available mineral nitrogen in soils. Denitrification is additionally enhanced in compacted soils.

Organic soils have significantly less mobile nitrogen concentrations and – thanks to the excellent soils structure – are more aerated. Both of these factors reduce $N_2O$ emission considerably. Permanent plant covers which are part of many organic cropping systems further reduce $N_2O$ emission. Methane emissions of organic rice and ruminant production equal those of conventional ones. In cattle breeding, longevity is a characteristic of organic systems. The ratio between the unproductive phase of young cattle and the productive phase of dairy cows is favourable in organic systems as far as methane emissions are concerned.

Soil erosion on agricultural land is accompanied by massive losses of carbon into the atmosphere. These losses occur in arable land (wind and water erosion) as well as on pastures (overgrazing through cattle, sheep and goats). Organically managed soils are significantly less prone to erosion and organic standards do not allow overgrazing as livestock charge is strictly limited per land area in respect to site specific productivity. Although many improvements in conventional arable systems have been discussed for decades, carbon loss is not slowing and, in fact, continues at high rates.

The carbon sequestration efficiency (tonnes $CO_2$-C per ha) of organic systems in temperate climates is almost double compared to conventional ones, when the total of above and below ground biomass of cash and catch crops and weeds is calculated. All farm comparisons and long-term field trials show significantly higher soil organic matter content when organically managed. In the Swiss long-term DOK experiment, in 28 years, the organic systems accumulated 12 percent and the bio-dynamic system accumulated 15 percent more carbon in the soil. The carbon sequestration rate was 575 to 700 kg carbon per ha per year. In the Rodale experiment in the USA, Pimentel, et al. reported an annual soil carbon increase of 981 kg per ha in the manure-based organic system and an increase of 574 kg per ha in the legume-based organic system. Carbon sequestration was on a perfectly comparable level in both the Swiss and USA long-term field experiments.

### Potential of Organic Agriculture for Better Adaptation to Climate Change

The Intergovernmental Panel on Climate Change defines adaptation to climate change as "adjustment in natural or human systems in response to actual or expected climatic stimuli or their effects, which moderates harm or exploits beneficial opportunities." This definition puts emphasis on the importance of farmer knowledge for adaptation to climate change, especially as its impact is very site specific (local) and all scenarios are very complex and unpredictable in many aspects. Therefore, Borron looked specifically into farm "community knowledge as a form of adaptive management."

Traditional skills and knowledge have been neglected in intensive agriculture although now they are being partly recaptured by integrated pest management. However, organic agriculture has always been based on practical farming skills, observation, personal experience and intuition. Knowledge replaces or reduces reliance on inputs. This knowledge is important for manipulating complex agro-ecosystems, for the breeding of locally adjusted seeds and livestock, and for the production of on-farm fertilisers (compost, manure, green manure) and inexpensive nature-derived pesticides. Tengo and Belfrages described such knowledge as a "reservoir of adaptations".

Agricultural production systems that emphasize resilience, resistance and site-specific adaptation to changing soil and water conditions, such as organic systems, are potentially wellpositioned to maintain production in the face of climate change.

Ewel describes soil organic matter as "the warehouse of most of the nitrogen, phosphorus, and sulphur potentially available to plants, is the main energy source for microorganisms and is a key determinant of soil structure." This underlines the importance of the findings concerning organic agriculture and soil fertility described in the previous section "Organic agriculture and stability and fertility of soils".

Farming practices such as organic agriculture that preserve soil fertility and maintain or even build up organic matter in soils are definitely well positioned to maintain productivity in case of drought, irregular rainfall events with floods, and rising temperature. Soils under organic management retain significantly more rainwater thanks to the "sponge properties" of organic matter. This was described for heavy loess soils in a temperate climate in Switzerland where water holding capacity was 20 to 40 percent higher in organically managed soils than in conventional ones. Pimentel et al., estimated the amount of water held in the organic plots of the Rodale experiment in the upper 15 cm of soil at 816 000 litres per ha. This water reservoir was likely the reason for higher yields of corn and soybean in dry years. Lotter, et al. found that water capture in organic plots was approximately 100 percent higher than in conventional plots during torrential rains. This reduced the risk of floods significantly, an effect that could be very important if organic agriculture were practiced on much bigger areas.

An additional strength of organic agriculture systems is their diversity at all levels including crop, fields, whole rotations, polycultures, farm activities (greater mix of different branches), and landscape on the farms. The high diversity of organic farms provides many ecological services that enhance farm resilience tremendously. Positive effects on pest prevention have been proven by Zehnder et al., Wyss, et al. and Pfiffner, et al. Similar effects of diversified agro-ecosystems on diseases and better utilisation of soil nutrients and water are likely to occur. Good quantitative data are not found in the literature, because such experiments are very difficult. However, it can be concluded that organic systems are very adaptive to climate change thanks to their high multi-level diversity.

## REFERENCES

El-Hage Scialabba, N.E. & Hattam, C., "Organic agriculture, environment and food security", Environment and Natural Resources Service Development Department, The Food and Agriculture Organization of the United Nations, Rome, 2002.

Fließbach, A. & Mäder, P., "Carbon source utilization by microbial communities in soils under organic and conventional farming practice", In: *Microbial Communities — Functional versus Structural Approaches*, Insam, H., Rangger, A. (eds.). pp 109-120. Springer, Berlin, 1997.

Forster, C.; Green, K.; Bleda, M.; Dewick, P.; Evans, B.; Flynn, A. & Mylan, J., "Environmental Impacts of Food Production and Consumption", *A report to the Department for Environment, Food and Rural Affairs*, Manchester Business School. DEFRA, London, 2006.

Hatfield J.L.; Sauer, T.J. & Krueger, J.H., "Managing soils to achieve greater water use efficiency: a review", *Agronomy Journal*, 2001.

Hole, D.G.; Perkins, A.J.; Wilson, J.D.; Alexander, I.H.; Grice, P.V. & Evans, A.D., "Does organic agriculture benefit biodiversity?", *Biological Conservation*, 2005.

Mäder, P.; Fließbach, A.; Dubois, D.; Gunst, L.; Fried, P. & Niggli, U., "Soil fertility and biodiversity in organic agriculture", *Science*, 2002.

Pimentel, D.; Hepperly, P.; Hanson, J.; Douds, D. & Seidel, R., "Environmental, energetic, and economic comparisons of organic and conventional farming systems", *BioScience*, 2005.

# 12

## Assessment of Organic Farming and Food Assertions

Organic farming owes its origins in part to the development of biodynamic farming. However this form of farming with its belief in cosmic forces has no place in any scientific discussion and is considered occult in character. The only resource that might constrain agriculture could be available farmland, but organic agriculture is less efficient in its use of land. Instead of lack of resources, it is now resource over-use, the global impact of their use along with global warming and a growing population that represent major concerns. Where appropriate these problems are referred to. A common claim, that agriculture has failed, is simply contradicted by the abundance of produce in present day supermarkets.

In a democratic society those who wish to farm organically (or biodynamically) or to eat organic food have a perfect right to do so. It would be helpful if the organic ideological community would recognise that right in reverse for others to farm in the way they wish. In a pluralist society all have rights.

Organic agriculture is defined by its regulations. Integrated Farm Management is defined by the LEAFUK audit, the main IFM group in the UK. LEAF species exacting standards of landscape, hedgerow maintenance, large field margins, quality soils and provides good animal welfare, all supposedly features only of organic farms in the public mind. But what exactly is conventional farming other than simply not being organic? And when comparisons are made, what evidence is there that the conventional farm is representative of this weakly characterised farming procedure in any way.

Unless evidence is provided on this point and perhaps the average conventional and average organic farm stipulated, published comparisons should be treated critically. Use of best practice conventional and organic farms (if that can also be defined) might improve the quality of comparison. The only specific features that really distinguish organic from all other forms of farming are a rejection of soluble minerals as fertiliser and synthetic pesticides in favour of natural ones. Any conventional farmer can choose to incorporate the supposedly unique organic environmental features into their farm and many clearly do; incorporation of manure to save on fertiliser, for example.

The variety of conventional farming procedures and farms causes difficulty in attempting comparisons. Not only do UK farms vary in climatic experience, ranging over nearly a 1000 miles of latitude, varying in rainfall, wind, average temperatures, exposure, slope, soil types and quality, mixed and arable, hedgerow structure, weed and pest problems but they also vary in managerial quality and attitude, level of farmer education, farmer understanding and finally wealth.

Although attempts are made to supposedly match soil quality when comparing conventional and organic farms, recent investigations using GPS controlled metre-by-metre surveys (precision farming) have indicated how variable is both the organic and mineral content of supposedly uniform fields. Comparisons that are made between conventional and organic farms commonly rely on farms nearby to each other to minimise (but not to eliminate) environmental variation. The presence (or absence) of trees and hedgerows which modify environmental impact are usually ignored. But managerial quality itself invariably fails to be investigated or estimated even though it is of obvious importance.

## FARM MANAGERIAL QUALITY

A farm is a densely interlinked structure forming a system or network that is ultimately controlled by the farmer and his degree of understanding of its behaviour (Fig. 1). In simpler terms we can recognise standards of farming excellence much as is commonly recognised in science, education and medicine. It is not difficult to recognise low farm standards; rusting machinery, half covered in grass; un-maintained or no hedges; holes in barn roofs, un-repaired gates; a pool of excrement outside the cowshed and so on.

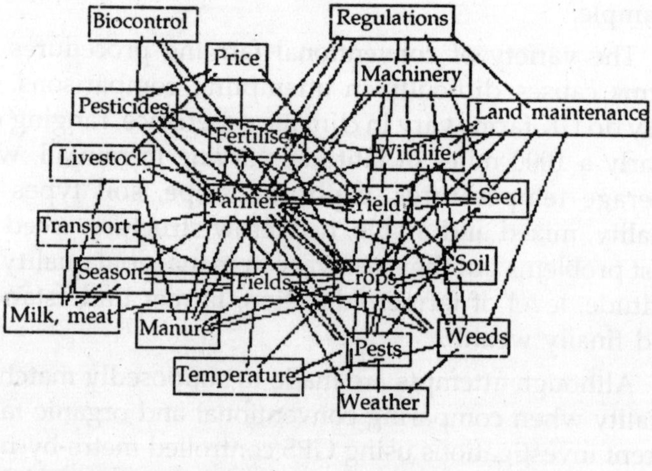


*Figure 1. A simplified view of a farm as a system.*

Furthermore, poor management and lack of biological understanding results in excessive use of costly pesticides thus wasting money; failure to use manure on a mixed farm to offset fertiliser costs wastes money, as well as failing to take advantage to maintain soil quality; poor treatment of animals leads to lower yields; failure to maintain hedgerows or large field margins limits pest predator numbers requiring more pesticide and so on. When farms fail the commonest reason is apparently poor financial, that is managerial, control in the face of a variable market. The IFM and organic benefits of wide margins, soil quality, landscape and good animal welfare help to limit the year-to-year variation in income, the major enemy of sustainable farm practice.

Perhaps the crucial feature here is that comparison of well-managed and poorly managed conventional farms will result in all the same landscape and animal welfare differences as those supposedly separating organic from conventional. The answer to this difficulty in scientific comparison of organic and conventional farming is simply to emphasise comparisons made on the same Fig. 1. A simplified view of a farm as a system. A system can be regarded as a network composed of inter-linked elements that influence each other. A boundary can be drawn around the system to simplify understanding.

To demonstrate linkages within the system it is sufficient to indicate that modification of one component modifies others in turn. In the figure these linkages are drawn as dotted lines. A proper systems structure would attempt to estimate the strength of linkages although bearing in mind that the strength will vary with circumstance (for example, manure will affect crop yield less if minerals have previously been added). But in a farm all system constituents are ultimately linked to the farmer and his ability to adaptively manage. Successful farmers appreciate the systems nature of their farm and act accordingly although they may do this intuitively and with wisdom gained from experience. The context around the boundary is composed of weather, crop price, price volatility and other financial arrangements that influence the overall farm system structure farm, matched fields and using the same farmer to avoid managerial variation in standard.

Most organic farms are supposed to remain small and self-sufficient whereas conventional farms are increasing in size the result of economic pressures. It will be increasingly difficult to make valid comparisons between the two. Farms, of necessity, do interact with each other and other facets of the commercial and food production community, constructing a complicated network. From what we now understand about network stability, small farms interacting minimally as a result of a policy of self-sufficiency do not generate stable networks, which instead require a limited number of hubs with very numerous interactive links (such as large conventional farms) coupled to many smaller constituents with fewer links. Organic regulations if expressed country-wide run directly counter to efforts to stabilise agricultural production. One possible structure that might be more stable is a diverse multi-hub network but this would see organic agriculture as merely a niche activity, not a primary form of agriculture.

Poor management survives because of subsidies, the curse of quality farming practice. Will organic action plans upgrade managerial standards then? Organic farming is recognised as higher risk because of crop failure and market variation. Thus only the best, most competent farmers are likely to try organic farming and probably only on a few of their fields. Therefore competent management will merely be transferred from one kind of farming to another with no overall improvement.

The simplest comparisons between organic and conventional agriculture is to use the same farm run by a farmer of excellent managerial quality. There are several established UK examples in the 10 year experiment at Boarded Barns, Ongar, Essex and the 7-year experiment at

Cooperative Wholesale Society (CWS) farms at Stoughton. Long-term experiments at Broadbalk, Woburn and Saxmundham may also be informative. There have been comparisons in other countries but climate and soil differences make these less useful in an article primarily directed to UK farming.

It takes decades to assess the effects of changes in agricultural practice and disadvantages of organic farming could take this length of time to appear. Many countries in the world practice organic farming now; but not by choice, but from poverty. When the UK was recognisably organic in the past, life expectancy was much lower, yields were lower, a substantial segment of the UK population were malnourished and/or ate a poor diet. Food in real terms was much more expensive. In 1900, for example, 60% of potential recruits were rejected for army service in the Boer war because of poor stature and weight resulting from poor diet. Price and the knock on effects on consumption are major concerns with organic food and one which should not be ignored. Food abundance reduces price.

### PESTICIDE CAUSES FOOD CANCER

Chemophobia, the unreasonable fear of chemicals, is a common public reaction to scienti.c or media reports suggesting that exposure to various environmental contaminants may pose a threat to health. The spectre of cancer birth defects and irreversible effects invariably scares people and various groups that campaign on environmental issues find the anxiety raised a useful source of political issue. Virtually all chemicals can be shown to be dangerous at high doses and this includes the thousands of natural chemicals that are consumed every day in food but most particularly in fruit and vegetables.

The assessment of chemical safety normally requires determination of the maximum tolerated dose (MTD) in rodents. Doses slightly lower than this figure are used to extrapolate linearly to a concentration at which observed effects would be limited to one in a million of the population and a "safe" exposure for the public set at figures 100-fold lower than this. Thus the safe level is frequently $10^{-5}$–$10^{-6}$ the MTD and on theoretical grounds it is very unlikely that the same cellular receptor (binding) sites could be occupied and the same effect induced, varying only in degree. Most cellular binding sites go from null saturation to full saturation over a 30–50-fold change in dose, not one million. The effects at high dose are therefore a priori expected to be qualitatively different from low dose and the effects of toxicity or cancer of little consequence to public exposure.

These natural pesticides are found in all fruit and vegetables; when tested at MTD they prove to be equally as damaging as synthetic pesticides. Furthermore the daily consumption of natural pesticides or carcinogens outweighs the traces of synthetic pesticides consumed by the public by many thousands to one. Mankind has always been exposed to "dangerous" chemicals and since many current crops have only recently been used as food and are also the result of extensive plant breeding, the kinds of natural chemical to which we are now exposed is too recent to allow for biological evolution to have ensured safe consumption. Solanine, chaconine, cucurbitacin, psoralen and genestein in potato, cucurbits, celery and soy have all been shown to have physiological effects if slightly elevated in food.

Carson alerted in her book to the potential effects of DDT accumulation through food chains, to observed eggshell thinning and thus brood failure of predatory birds. Attempts to repeat these observations in the laboratory failed. However because organochlorines like DDT are environmentally stable, bans on its use were instituted in the early 70's. In western countries DDT was replaced by effective but far less stable pesticides. But pressure on developing countries to do the same led to enormous increases in malaria, a disease that particularly kills young children.

In developing countries, cheap DDT also killed many crop and other human disease insect vectors and lowered food prices that benefited the poorest most. Recent evidence has revealed that organochlorines are formed in substantial amounts (mg/kg) in decaying plant materials such as crop roots in soil. Because organochlorines are fat soluble, they will be found in all plant food materials and mankind has therefore been exposed to these stable chemicals probably throughout evolutionary history and so far as is known without effect. In fact any natural pesticide can bio-concentrate like DDT if it is fat soluble. Potato contains the fat soluble neurotoxins, solanine and chaconine, and high doses of these have been shown to cause birth defects.

However chemophobia is the commonest reason for the public to buy organic food on the assumption that such food is free of synthetic pesticides. Organic food contains synthetic pesticide traces although understandably the amounts are lower than in conventional produce.

Life expectancy continues to increase unabated and figures specific for the UK are to be found in Lomborg. Centenarians are now ten times more common in the UK than they were 50 years ago. There has been no obvious effect of the introduction of pesticides and use through 50 years on life expectancy or on general public health.

### Agricultural Exposure to Pesticides and Cancer

Specific synthetic compounds have been linked to cancer as a result of occasional high level of occupational exposure or accidents such as Bhopal. In order to link together public exposure and cancer there must be correlations between the level of exposure and the magnitude of the response, consistent results from a number of different studies, and biological plausibility based on studies in laboratory animals. If there is such a relationship, it can be hypothesised that it should be most readily apparent between cancer rates and those most exposed to pesticides.

Since farmers, foresters, pesticide users and manufacturers are by occupation more likely to be exposed to higher pesticide hazards than the general public, many published studies have investigated cancer rates in these groups usually through cohort investigations using matched controls from the public particularly in age and social status. Of 12 separate investigations on farmers involving in total about 300,000 people, 11 found that farmers had overall cancer rates very substantially lower than the general public. Slightly smaller numbers of investigations on pesticide users and foresters revealed a similar (8 out of 11) trend whilst only with manufacturers were cancer rates similar to those of the members of the public.

The reasons why farming is so healthy are not known but these data indicate not only a null result for the hypothesis relating pesticide exposure to cancer but a consistent result for the alternative, that pesticide exposure may protect against cancer. These results are also consistent with a hormetic effect of low pesticide doses. Farmers do have higher rates of the less common lymphoma but evidence indicates that lymphoma is unrelated to pesticide exposure. Higher exposure to animal disease viruses or fungal disease may be a more plausible hazard.

Cancer death rates in the UK are in decline particularly when lung cancer induced by smoking is removed from the 50 years of detailed age-related statistics published by the epidemiologists Coggon and Inskip. They conclude "there is no evidence that pesticides have had a major impact on overall rates of cancer". Cancer rates increase over 100-fold between the ages of 20–70. Thus increased numbers of cancer cases accompany an increasingly healthy population dominated by elderly people. Only when cancer rates (cases/ numbers of people) are expressed for each defined age group as shown in Coggon and Inskip for every 5 years of age are genuine trends detectable.

Although cancer death or cancer incidence, could be examined, it was common 50 years ago for pathologists to report that coronary fatalities also had advanced cancer. As fatalities diminish from heart failure more die from cancer. The estimates of human recovery from cancer do not seem to be above 5–10% of diagnosed cancers, so although incidence might be considered a more useful criterion of environmentally damaging materials, death rates may be considered a more valid, easily diagnosed measurement. It is among the young and middle aged that anything untoward concerning illness first appears and these two groups show the greatest declines in cancer rates over the last 50 years.

Some cancers such as breast cancer are increasing but the factors that conspire such as lifetime oestrogen exposure and most significantly obesity are the probable major risks. The stomach is the most likely tissue substantially exposed to ingested pesticides but stomach cancer rates have declined by about 60% in the last 50 years. The Government's Committee on Carcinogenicity after examining the evidence in detail concluded that breast cancer rates are completely unrelated to organochlorine exposure. Perhaps more intriguing with the large investigations involving 50–150 thousand women, cancerous tissue has slightly less DDT, DDE the main metabolite and PCB's although the difference is within the standard error. The issue of DDT traces is primarily concern about bio-accumulation rather than known toxicity.

*Possible xeno-oestrogenic effects of pesticides*

During the last decade there has been debate about a

potential new threat to human health, exposure to chemicals potentially with endocrinologic activity; the so-called "endocrine disrupter" hypothesis. However measurements place this argument in perspective. The major dietary source of oestrogen-mimicking molecules are flavanoids and isoflavones in fruit and vegetables. The daily consumption, about 1–2 g/ day, represents oestrogen equivalents about 3% normal circulating level of oestrogen in the non-menopausal female. The current dietary exposure of dibutylphthalate from plastic would increase circulating oestrogen-mimic content by 0.0006% and from dioxin and organochlorines by $6\times10^{-6}$%. In contrast the levels of oestrogen mimics in certain alternative medicines is sufficient to cause physiological changes in treated men.

Concern over the consumption of soy genestein particularly in vegetarian diets using soy milk for prepuberty children has been raised. Genestein is an established genotoxin and evidence from human studies suggests potential modifications of thyroid function leading in the long term to goitre.

*A diet high in fruits and vegetables cuts cancer rates in half*

Block et al. summarise some 200 investigations on diet and cancer using cohort studies over the previous 20–30 years. These authors indicate that for virtually all the major cancers, a diet high in fruit and vegetables cuts cancer rates approximately in half. These investigations involved very large numbers of people, only in western countries, and those under investigation consumed conventional produce containing the inevitable pesticide traces. The fact that increased consumption of these supposedly damaging substances in fruit and vegetables actually makes you healthier, contradicts any simple assumptions concerning pesticide traces and cancer.

Furthermore we now know that fruits and vegetables contain an estimated 10,000 secondary products. When tested in exactly the same way as synthetic pesticides, at the MTD, 60% of each were carcinogenic. But a number of these like limonene, perillyl alcohol, quercetin, allyl isothiocyanate, all carcinogenic in high doses are now used to treat cancer but at low doses. These natural pesticides (so-called because they are synthesised by plants to poison insects and herbivores) will also be consumed in greater amounts with diets high in fruit and vegetables. The quantity on average consumed/day of natural pesticides outweighs synthetic pesticide residues by 20,000 to 1. These results indicate that the tests and assumptions, used to identify synthetic pesticide traces as damaging to humans, lack meaningful interpretation since consumption of natural pesticides does not induce cancer.

***Hormesis indicates the potential value of synthetic chemicals in the diet***

Hormesis is the paradoxical effect of toxic chemicals at low concentration. Many toxins at high concentration are likely to have the opposite (and thus probably beneficial) effect at low concentrations. There is an enormous toxicological literature on this subject and Calabrese and Baldwin refer to hormesis as a revolution in toxicology.

The standard method of assessing toxicity was based on a model of cancer that is now outdated. This model assumed that toxic at high concentration simply meant less toxic at low and that there was no threshold below which effects of pesticides were absent. Consequently no concentration of any toxin could be considered absolutely safe. The theory became known as the linear nothreshold theory (LNT) and was formulated in the 1970s. Developments in our understanding of cancer have shown the basic assumptions (all mechanisms of cancer the same; cancer as a monotonic disease) that underpinned the model are incorrect.

Instead many chemicals in the low dose region have a dose response curve that is J-shaped, indicating, for example and where relevant, that low toxin consumption actually inverts the physiological effects of high dose. Data collected from 5000 dose response measurements indicate that low doses of many supposedly toxic chemicals, metals, pesticides, fungicides, petroleum fractions, radiation and even diluted factory effluent either reduce cancer rates below controls, increase longevity or growth in relevant organisms.

There are many familiar examples of hormesis. Aspirin, where one or two tablets/day improve the circulation, but 30 will stop it altogether. Virtually all pharmaceuticals are similar. Fluoride to strengthen bones and teeth but fatal in large doses. Antibiotics and vaccines where a 10-fold higher than recommended dose can kill. Sunshine induces vitamin D formation in small doses, but melanoma in large doses. Radon, a radioactive gas formed in higher amounts in granite houses but those who live in such houses have lower rates of lung cancer. Vitamin A is necessary for vision but teratogenic at slightly higher than recommended doses. Insulin is the crime writer's invisible murder weapon at high dose, but necessary to control human blood glucose level at low dose. Drinking water suddenly in very large amounts has been known to kill by upsetting the ion balance.

Hormesis is thought to result from the beneficial effects of mild chemical stress at low concentration potentiating the immune and oxidative defence systems; in turn destroying pre-neoplastic loci. Well-established cases show that low doses of chemicals like cadmium or dioxins or anthracenes, for example, actually lower cancer rates below control levels. In other words if synthetic pesticide traces are at a level to have any effect at all, the effect is as likely beneficial as anything. The expected hormetic effect of natural pesticides in fruit and vegetables potentially accounts for their beneficial action in reducing cancer.

### *Price of fruit and vegetables determines health*

Organic food is more expensive because it uses land less efficiently, as will be seen later. However only 21% of the UK and only 20% of US citizens actually eat the necessary two fruit and three vegetable portions/day recommended by respective governments to provide minimal protection against cancer. Those numbers will not be increased by raising the price of produce because it is known that price is a very strong determinant of consumption. Furthermore raising the price will not encourage the cultural change necessary to increase fruit and vegetable consumption. For those on low incomes, if organic food is purchased less will be consumed, with consequences in cancer rates decades later.

### *Where organic naturalism ideology leads*

A common claim is that intensive agriculture seeks to dominate nature whereas organic agriculture works with nature. It is not very clear what the statement means since no form of farming is natural relying in every case on clearing ground and destroying the natural environment (forest and the associated organisms) that was there before.

However the phrase uses the term "nature", appealing to a common phenomenon described by E.O. Wilson as Biophilia. Human beings recognise the natural world and often feel a strong kinship with it because we are living organisms too. Simplistically this turns into "natural" equated with "good" and human activity as "synthetic" equated with "bad". This view is self evidently, a version of original sin. However child death, starvation, plague, cholera or even SARS are all the results of natural activities; and in contrast, synthetic activities, such as architecture, painting, music and literature are all man-made.

The limitations of this view can be exposed by comparing a natural pesticide with its synthetic equivalent.

Table 1 makes some limited comparisons between mancozeb, a synthetic copper fungicide usually used to treat late blight, and the organic pesticide equivalent, copper sulphate. The full table can be found in Leake. In environmental qualities, mancozeb is superior in all categories compared to copper sulphate. In terms of human health, copper sulphate is corrosive and toxic and has caused liver disease in European vineyard workers. Although the EC theoretically banned copper sulphate in 2002, no alternative has been found for organic farmers and thus it continues to be used.

*Table 1. A comparison of the human and ecotoxicity of mancozeb and copper sulphate*

| | *Mancozeb* | *Copper* |
|---|---|---|
| *Human health* | | |
| LD50 | >5000 mg/kg | 50 mg/kg |
| EPA class | Practically non-toxic | Corrosive and toxic |
| Health effects | Non-toxic by oral route | Kidney and liver damage |
| *Ecotoxity* | | |
| Earthworms | Low toxicity | Very toxic |
| Birds | Low | Moderately toxic |
| Small mammals | Non-toxic | Harmful |
| DT50soil | 6–15 days | Non-degradable |

The consequences of not using copper sulphate properly have been reported as organic farms acting as repositories of late blight, a serious disease of potato or seriously damaged orchards. Any sensible approach would determine use based on toxicity. Other organic pesticides such as soap hardly merits the term natural. Rotenone, another organic pesticide, has been recently associated with Parkinsons disease, and another, Bt spores, used to kill insects has been associated with damaging respiratory effects. Synthetic pyrethroids can be as unstable as the organic pesticide, pyrethrum, but work at much lower concentrations. How natural are the organic pesticides sulphur or oil when they have to be mined, chemically modified or distilled before use?Agriculture tends in one sense to mimic the effects of individual tree loss in forests. When a mature tree falls over, light penetrates the forest floor, there is flush of nitrate and water in the soil. Weed seeds, which are sensitive to light and/or nitrate, germinate.

Most crops are derived from weeds although their germination requirements for nitrate and light have been bred out. Farmers mimic tree loss by clearing ground. The assumption that nitrate is somehow an unnatural fertiliser is contradicted by this perfectly natural process. But the farming activity most deserving the term natural is no-till agriculture which mimics in part the annual cycle of the meadow or prairie.

## ORGANIC AND CONVENTIONAL SOIL PROPERTIES

The development of thin film technology several decades ago for growing greenhouse crops demonstrated that soil is not actually essential for good crop growth and that only minerals are actually required for active and healthy plant growth. Aeroponics using misted mineral solutions generates the most abundant and healthy root systems. Nevertheless most crops will continue to be grown in soil.

### Low Organic Yield more Natural

From many reports it is clear that organic yields are usually lower, the extent depending on the crop. Leake at CWS agriculture using fields at the same farm reported that organic wheat, beans and peas yields were 60–70% whereas oats were 85% conventional yields (Table 2). Such measurements have been repeated on numerous occasions with variable results. Usually organic yields are much lower but occasionally they can match conventional productivity in a single year. But most organic farming involves a ley period in which clover or alfafa and grass is grown to allow nitrogen fixation and provide the soil with nitrogen once ploughed in. No organic yield is obtained throughout the ley period unless it is grassed and used for cattle, whereas in a conventional farm the same field can be in use continuously for arable crops.

Many organic farmers use a field twice for arable crops after the ley period and since in the second year less nitrogen is available, the yield further diminishes. Thus occasional claims that yields can be the same as a conventional farm may only be true one year out of three. It is important to compare total yields over a number of continuous years, not on a single year. Thus Drinkwater et al. claimed that organic and conventional corn yields were identical without clarifying the necessity for twice as much land to achieve that yield. The Boarded Barns study routinely reported organic wheat yields using animal manure of about 50% conventional production. Winter wheat is still the staple of most UK arable farming.

*Table 2. Typical yields of the major crops grown conventionally and on Soil Association licensed organic fields on the same farm and using the same farmer*

| | *Winter wheat* | *Winter oats* | *Winter beans* | *Dried peas* |
|---|---|---|---|---|
| Organic | 4.82 | 4.87 | 2.60 | 2.13 |
| Conv.7.12 | 5.75 | 3.60 | 3.47 | |

Such yield differences indicate that organic farming uses good farmland less efficiently. Instead an equivalent yield to organic farming could be achieved on less conventional farm land and the remainder (20–50% of the farm) more profitably employed as woodland, or in willow plantations for bio-fuel, (thus recycling carbon) or returning it to other natural conditions such as fenland. These considerations of efficiency are much more important overseas where growing populations using inefficient organic agriculture, will simply cut down more tropical forest than they need to feed the growing population. Currently, for example, Mexican peasants destroy 3 million acres of virgin tropic forest/year to slash-andburn agriculture.

However the UK government has signed up to a protocol to ensure that 10% of all fuels used in the UK by 2010 come from natural sources such as ethanol, thus ensuring that carbon is recycled. On this basis wheat is likely the source for fermentative ethanol production and increasingly, intensive agriculture the likely means for its production.

One reason has to do with the prohibition on the use of very soluble fertiliser. There is a major surge in crop growth suddenly in late spring and mainly in leaf production. Leaves are the most nitrogen-rich tissue in a higher plant and consequently there is a relatively sudden heavy requirement for nitrate to produce leaf protein for chloroplasts and photosynthesis. The vegetative reserves laid down during canopy expansion help provision the seeds when they form. Therefore maximal seed yields are likely to be obtained only when the provision of soil nitrate and the associated crop requirements for leaf production are synchronised. This temporally uneven requirement for N in springtime only can be matched by careful application of very soluble fertiliser.

However a process that ploughs in material which is only slowly degraded over many months or even years cannot release minerals in the short intense burst as required for plant growth. Provision of nitrogen by decay of organic material (mineralisation) throughout the season produces nitrate when it is little needed and does not provide sufficient when it is. Furthermore mineralisation is greatly speeded up when the field is ploughed, months ahead of the time required by the crop.

In a detailed examination of mineral availability, Berry et al. determined the grain N as a % in wheat, the grain yield and inorganic fertiliser applied (Fig. 2). Their comment is that although the amount of N in organic soils should be equivalent to 300 kgN/ha based on soil analysis, organic wheat plants act as though there is only about 50 kgN/ha available for growth and seed formation. Berry et al. also indicate that the common practice of applying manure or slurries to ley legumes simply diminishes the amount of N fixed by the legumes resulting in a waste of manure. More crucially these measurements indicate that overall analysis of N on organic soils is misleading when such a mismatch between unavailable and available N is so clear. It also suggests that excess N in organic soils, as reported by Watson et al., appears at the wrong time of year and is likely to disappear into waterways and courses since it remains unused by the crop. Since organic regulations stipulate the use only of supposed natural minerals that decay slowly and are therefore weakly soluble, general mineral analysis might well prove entirely misleading as to how much mineral such as phosphate is actually available to crop plants during the current season.

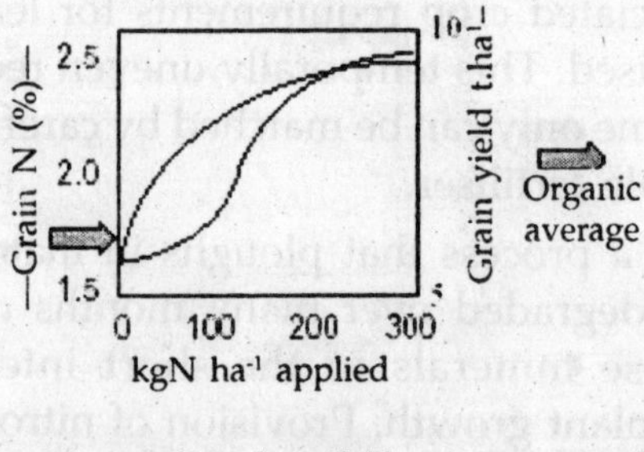


*Figure 2. Does the supply of available nitrogen restrict organic productivity?*

Ecologists regard fitness as usually measurable in terms of seed yield. On this ecological basis, conventional farming produces fitter plants than organic agriculture.

### Soil Fertility on Organic Farms

One of the claims of organic agriculture has been the supposed benefits of organic regulations to soil fertility and soil structure. A series of reviews on organic soils constructed by organic and conventional researchers and published in Soil Use and Management 2002 clarifies the situation. The concluding and summarising chapter states that "Soil fertility is defined as an ability of a soil to provide the conditions required for plant growth", a view echoed by Russell and Voelcker much earlier. Given the complexity of soil attributes which modify plant growth and yield and with each crop having different optimal soil requirements, it is doubtful if a more precise definition can be constructed.

Critically, crop plants interrogate the soil in which they live and then respond accordingly. In the wild, many types of soil are occupied by individual species, and plants have developed necessary control systems to mitigate some of the variations they come across. If conventional soils were poor from a plant point of view, this would result in poor yields and clearly they are not.

After a detailed consideration of soil organic matter, minerals and nutrients in labile mineral pools in soils, Stockdale et al. also conclude that "Although nutrient management in organically managed soils is fundamentally different to soils managed conventionally, the underlying processes supporting soil fertility are not". But ploughing in of legume rotations in organic farming provides substantial amounts of soluble minerals to the soil since the vacuole of each cell of a mature plant contains mM concentrations of soluble calcium, magnesium, phosphate and 2–300mM potassium chloride. Therefore the nutrient management processes are not completely different.

### Produce Better Organic Soil Structure

The assumption is that organic farms using organic

material added back to the soil must produce a better soil. It should be added that the criteria used to define a good soil are those of the soil scientist, not that of the individual crop plant which may have a very different perspective.

Surprisingly few differences in organic matter content were found between organic and conventionally managed pasture soil, and only trivial differences in 20% of the soils between conventionally managed and organically managed arable soils were detected, usually it seems on stockless conventional farms where differences might be more likely. There is a large background of soil organic matter regardless of farming system deriving from crop roots and root exudates. No consideration seems to have been given in the reviewed literature to the need for careful matching of managerial standard in the comparisons made by Shepherd et al.

But best-practice, mixed IFM farms will equally apply manure to the soil when they have it and in this case no meaningful difference between IFM and organic soil is likely. The literature is ambiguous and what differences emerge can usually be accounted for by different rotations and different tillage practices. But tillage practice is a choice made by the farmer and a conventional farmer can as easily add organic material to the soil as an organic one. There is no need to change agriculture radically to ensure that this is done; merely pointing to the saving in fertiliser costs should provide sufficient impetus. Managerial quality or choice, not farming mode, is the determinant of soil quality.

### Microbial Diversity of Organic Soils

If there is greater amounts of organic material in the soil it is usually assumed that this will be accompanied by higher microbial numbers since most soil microbes are heterotrophs requiring organic compounds for growth. Whether higher microbial numbers are actually beneficial or merely asserted to be beneficial seems unclear. But the evidence on microbial populations is conflicting and ambiguous and no clear picture emerges as to whether there are distinct differences in the size, composition, biodiversity and activity of the soil microbial biomass attributable to farming management practice.

Very large number of microbes are only required if organic material has to be broken down to provide minerals for plant growth. An inability of organic farming to provide sufficient N at the time required suggests there are probably, in relative terms, insufficient microbe numbers.

## Conventional Farms Pollute Waterways with Nitrate and Organic Farms Do Not?

Two substantive investigations indicate that nitrate pollution from organic and conventional farms are effectively identical. The first of these reported nitrate leaching through two ley periods followed by three arable periods and showed that when averaged over the 5-year period, nitrateleached losses were identical at about 50 kg/ha. Stopes et al. found that "nitrate losses following arable crops averaged 47 and 58 kgN/ha for organic and conventional fields, respectively, following arable crops with part of the difference being due to the greater proportion of non-cereal break crops in the latter". Clearly there were difficulties in matching farm management. These figures are little different to those described by Pickett and Goulding. Nitrate losses during the organic ley phase were similar (45 kg/ha) to those from conventional long-term grass receiving fertiliser levels below 200 kgN/ha (44 kg/ha).

## Do Soluble Minerals Damage the Soil?

The Broadbalk experiment has now lasted 160 years being started in 1843 by Lawes and Gilbert. In two sections of this experiment, winter wheat has been grown continuously on land either given only manure (35 tonnes/ha) or only mineral fertilisers. This far-sighted experiment, still of direct relevance to sustainable agriculture concerns, has shown that the yields from the two treatments have remained identical for 160 years. Yields have changed in parallel in the two sections with the introduction of new winter wheat varieties, herbicides, fungicides and all the constantly improving agronomies that are now part of farming. It is quite clear from these data that, within the confines of the experiment, conventional agriculture using minerals is sustainable; 160 years is long enough. The soil carbon content has increased in the manure-only treatment by more than two-fold but this has not enabled a higher yield to be obtained. The soil carbon content has increased in the mineralsonly treatment by about 20%, the result of leaving root material in the soil.

Whether such equivalence in yields over 160 years would be obtained with other crops is not known but the experiment makes the critical point; that soluble mineral-only treatments do not damage the soil from the point of view of the crop plant. However no current farmer in the UK would use a system without rotation and therefore this aspect of the Broadbalk experiment does not now accurately mimic either conventional or organic farming. Pests are better dealt with by rotating crop use. Furthermore identical yields between the manure-only and mineral-only section were obtained by using very large amounts of manure (35 tonnes/ha). But the consequence was that nitrate run-off from the manured section was much higher than the mineral-only section.

The question of sustainability was also examined with other long-term 50-year experiments at Woburn and Saxmundham which commenced in the 19th century and is described in detail by Johnston and by Russell and Voelcker. These experimental set-ups used either single continuous winter wheat or the Norfolk four course rotation. But similar results were obtained; use of mineral salts-only compared to manure-only did not diminish yield in relative terms.

### Is Mineral Recycling on Organic Farms Sustainable?

A well grown crop of potatoes, (60tonnes/ha) for example, if sold, removes from the farm 29, 338,12, 4 and 5 kg/ha of P, K, Mg, Ca, Na in the tubers which must be replaced otherwise the soil is mined and becomes seriously deficient as are found in many present day African soils. There are currently two primary sources permitted by UKROFS (UK organic regulations).

1. Organic regulations list mined $CaCO_3$ (chalk), KCl (sylvanite), $MgSO_4$ (kaiserite), rock phosphate, trace elements and eight other non-renewable inorganic chemicals, all for a claimed chemical-free agriculture. Such materials have to be mined and transported but the energy costs of transport are usually ignored in energy assessments even though rock phosphate originates in Africa. The lack of purity of such crude mineral sources does not seem to concern organic regulatory bodies. Rock phosphate allowed under organic regulations contains much higher levels of cadmium than purified conventional P. There is concern over the potential deficiency of micro-nutrients in organic food such as selenium. There are no stated selenium deficiency symptoms for crops but serious consequences for selenium deficiency for human health. Recycling on the farm itself, the supposed goal of organic farming, will not amend this situation.
2. The second source is manure and/or hay from intensive agriculture. Although organic regulations stipulate that such products should come from extensive agriculture, this conflicts with a requirement to balance the number of animals on every farm to the manure requirements so that no excess organic manure

is produced. In this situation organic agriculture becomes dependent on intensive farming, a form of farming it considers unsustainable. But in the absence of conventional farming, organic as presently constructed will be completely unsustainable too.

3. A further source of minerals is purchased animal feed. But this again is produced from conventional agriculture using minerals. However organic regulations would not permit this source of minerals in animal feed for general use leading again to questions over sustainability.

Conventionally generated manure uses soluble minerals for the production of animal feed in the form of wheat or corn. The energy costs of producing minerals for animal feed (often commented on) are simply carried through to the energy costs of producing the same amount of K, N or P in manure (often ignored). Recent detailed analyses of many field soils on a metre-by-metre basis indicate substantial variability in mineral content and yield of individual fields. Because dumping manure requires more labour than distributing minerals, it has been found that fields on some organic farms farthest from manure heaps are the least treated and experience most deficiency. Congregation of animals causes further uneven distribution which can last years. Such maldistributions make calculation of available N on organic farms very uncertain which overall assessments of the whole farm do not mitigate.

Watson et al. in an analysis and review of organic farm budgets and mineral recycling conclude "The data presented here also suggest some cause for concern in relation to sustainability of organic dairy systems because of their dependence on imported feedstuffs and bedding for P and K and for N on the very variable fixation by legumes of imports of manures or composts". On its own then, organic farming is not sustainable except as part of another form of agriculture on which it is actively dependent and which it opposes. Either the lack of sustainability will have to be recognised or organic will have to import minerals of one sort of another which due to slow breakdown will cause problems in matching requirements with yield as figures already indicate. In that case organic farming will no longer be able to claim to be chemical free. The notion that organic farms can operate as small, totally self-contained entities is simply a delusion.

**Is Manure Better than Minerals?**

Analysis of manure indicates the presence of extensive amounts of the soluble minerals (such as K) which are prohibited under organic regulations. But these minerals will simply wash into the soil and be easily lost unless applied to a growing crop. Potassium chloride (sulphate, phosphate or nitrate) is no less soluble in manure than it is in soluble fertiliser form. Loss of N in ammonia from fresh manure represents waste and at least half of the manure N can be lost during composting from volatilisation. But composting is essential to eliminate dangerous organisms. It is a common assumption that in some way manure contains no chemicals and that therefore organic farming is somehow chemical free. But fresh manure is known to contain mercaptans, aliphatic aldehydes, organic acids and lipids, phenols and mcresol, various amines etc. A full analysis of all the chemicals in manure seems currently to be unavailable but based on previous experience at least half of these organic chemicals are carcinogenic in rodent tests. The major constituents of manure are lignin and cellulose breakdown products. Overcash et al. and Eghball and Power provide detailed analyses of different manures indicating that all the major (N, P, K) and some of the minor minerals can vary 10–30-fold in different manures. Such variation is a further cause for concern, matching those expressed above for using only legumes for nitrogen production.

It is well known that farm animals shed dangerous organisms from their gut and cases of *E. coli* 0157 and Salmonella poisoning from manure contamination are well established. Although composting manure can destroy microorganisms, the process is arbitrary and uncertain as it is dependent on the shape, volume, position and time of composting. Killing dangerous organisms in manure heaps is dependent on critical temperature (usually 60°C) being reached and maintained for 90 days. There is no simple way of determining this temperature/time requirement and temperatures at the compost surface are usually lower than inside.

The destruction of food-poisoning viruses (such as the Norwalk-like viruses) has not been examined at all although such incidences of virusinduced poisoning are suggested to be increasing. Child death from eating organic parsley has been reported in the medical literature. While problems with manure apply equally to conventional mixed farms as well, a conventional farmer can use manure just as a soil conditioner and provider of organic material to improve soil crumb structure. Such farmers are therefore not under the same pressure as an organic farmer who needs to apply manure quickly to the soil before much of the N disappears into the atmosphere as ammonia and probably nitrous oxide. Neither organic nor conventionally grown foods are free from pesticides but the health risks are considered to be higher from food borne pathogens than pesticide residues.

## Is Organic Food More Nutritious, Healthier and Tastier than Conventional Products?

Many taste assessments of organic and conventional foods have been properly made. In these the individual is provided with three pieces of the same produce and asked to state which two are the same. Such tests have indicated that there is common confusion of freshness for organic produce and that the public cannot easily distinguish organic from conventional.

Although it has been claimed that organic produce is healthier food than conventional produce, the current evidence does not support this contention. Most papers indicate that the composition of organic produce is not significantly different from conventional food. Mineral composition of conventional food can vary very substantially, dependent on conditions of growth. Summaries from nutritionists and others indicate that occasionally there may be slight increases in vitamin C although the figures are not particularly consistent and are strongly biased by two isolated measurements by Schupan many years ago in lettuce and cabbage. However vitamin C may accumulate when oxidative stress is experienced, a consequence of disease and perhaps deficiency. Mutation rates and DNA rearrangements may also increase as a result of stress and disease, as has been observed.

There is no indication however that the UK public is in anyway short of vitamin C; in fact current consumption levels of 70–90 mg/day are close to the renal plasma clearance value of about 100 mg/day. WHO recommendations are for 45 mg/ day. At least 30 chemicals or additives are allowed and used for organic food processing, a fact not commonly known. Bourn and Prescott conclude that "with the possible exception of lower nitrate levels there is no strong evidence that organic and conventional food differ in their concentration of various nutrients. While there are findings indicating that organic and conventional fruits and vegetables may differ in a variety of sensory qualities, the findings are inconsistent." They suggest that freshness is an important criterion often confused with organic food because of its local sourcing.

Williams, a human nutritionist, concluded "there appears to be a wide-spread perception amongst consumers that such organic methods result in foods of higher nutritional quality. The present review concludes that evidence that can support or refute such perception is not available in the scienti.c literature." Occasional attempts have been made to compare health outcomes between those only consuming organic or conventional produce " Comparison of health outcomes in populations that habitually consume organically or conventionally produced foods are flawed by the large number of confounding factors that might contribute to the differences reported."

Different conditions of cultivation could well give rise to pronounced variation in the chemistry of the food produced and, in particular, Williams suggests that stress for organic plants might well be a problem, inducing the formation of a variety of chemicals whose total contribution to human health is simply unknown. Given the wide variety of ways in which even organic farms are run it is unlikely that any standard guarantee for food quality can be produced.

Organic is a mechanism or process; it does not guarantee the product. Managerial quality seems equally important. It needs emphasising in conclusion that if the composition of organic food is different, particularly in secondary products, there can be no guarantee of long term safety because human disease and mortality statistics to justify the safety of organic food are not available. In contrast substantive information on the consumption of conventional fruit and vegetables and such statistics are present in abundance.

### Is Organic Farming Environmentally Superior?

It is a common assumption that organic farming is better for the environment; but better than what? It depends on what is considered the most valuable criterion of environmental benefit and that is not easy to define.

### Measurements on the Same Farm

The Boarded Barns study reported that 80–85% of biodiversity on any farm existed in the field margins and hedgerows. Measurements covered small mammals, collembola, (mites that damage pests), micro and predatory arthropods, predatory beetles and birds. What was in the cropped area was of little overall significance for biodiversity. Thus any farming mode that maintains margins and hedgerows is likely to be as good as any organic field and the effects of pesticide application on the cropped area of little significance. In the Boarded Barns study, small mammal activity examined wood mice, bank voles, common shrews. Based on trapping, activity and diversity was identical between organic, conventional and IFM farming. In Collembola the only significant change was a higher proportion of individuals on IFM soils, conventional and organic being the same.

Species richness was also similar. Earthworm densities were highest in the no-till IFM fields with organic second at about twice that of conventional fields. Numbers of earthworm species were however similar. Microarthropods were five-fold higher on IFM fields than organic or conventional fields. The numbers of predatory arthropods, in particular carabid beetles, were however higher (about 25%) on organic soils. But the crop in the field played a greater part in carabid variation than the mode of agriculture, being four-fold higher in bean crops than wheat. In these measurements organic wheat had only one half the numbers of carabids than conventional wheat.

### Measurements Comparing Different Farms

Similar reports of higher numbers of carabid beetles on organic fields have been reported in a number of investigations. For example Shah et al. found three-fold higher numbers of carabids but almost all was due to much increased numbers of only one species of beetle, Pterostichus melanarius, the common black beetle which also eats earthworms. However Armstrong detected no differences in potato crops suggesting crop variety may be more important than mode of agriculture as was found in the Boarded Barns study.

Ground beetles are much easier to capture which accounts for their popularity in measurement, but using vacuum collection other kinds can also be surveyed but the measurements are few in number. The numbers of Staphylinids, (winged predatory beetles), Coccinellids (lady birds) and Cantharidae were very substantially higher in conventional fields as were all other measured beetle classes apart from Carabidae. Moreby et al. indicated that some of their measurements comparing organic and conventional fields were made on the same farm and thus would satisfy the requirement of managerial equivalence.

Feber et al. compared butterfly numbers on organic and conventional farms and found more on organic farms of the non-pest variety although differences varied with the crop and the year. Thus the authors concluded that crop patterns not mode of agriculture might be the over-riding determinant. Perhaps this is similar to the crop effects on carabids described above. No comparisons were made with IFM farms and since field margins and hedgerows are similar between organic and IFM farms, butterfly numbers relying on these undisturbed parts of the farm are unlikely to be different.

### Biodiversity Indices of Invertebrates on Organic and Conventional Fields

Whether having more of one particular species of beetle makes the ecosystem more or less stable is very unclear. However application of measures of biodiversity (the Shannon-Weaver index for example) in Shah et al. indicated the highest beetle biodiversity to be on conventional fields.

All others have reported the same outcome using biodiversity measures. That should not be a surprise, there is a more uniform distribution of numbers between different beetle species on the conventional fields.

Greater diversity tends to be found in stressful situations with pesticides providing the stress. If diversity is related to stability then the conventional field may be classed as more ecologically stable. Foster et al. reported that water beetle fauna of higher quality and quantity were found now in 1990 than in 1938 when synthetic insecticides were not in use. Water beetles are sometimes used as indicants of water quality.

Deep ploughing can disrupt the carabid community, and no-till methods will leave the highest numbers with reported increases anywhere up to 20-fold down to zero. Hedgerows, beetle banks and conservation headlands, all part of IFM increase carabid beetle numbers.

Other insects were also surveyed by Moreby et al. They found higher densities of nematocerean and acalypteran diptera, hemiptera (especially aphids), aphid specific predators, parasitic hymenoptera as well as most beetle fauna in conventional fields. Higher densities of weevils, spiders, springtails, plant hoppers and saw fly larvae were found in organic fields. Moreby et al. conclude that "although some significant differences in density of some arthropods were found in organic fields of winter wheat, one of the most striking observations was the lack of large scale treatment differences that were expected from such fundamental differences in the use of pesticides and different approaches to crop rotation and soil cultivation".

### Bird Populations

Declines in numbers of particular bird species normally found on both UK farmland and gardens have been reported in the last 30 years and intensive agriculture has been blamed by some for this situation. However the situation is complex and to claim that it is just changes in farmland management is simplistic. Declines in birds that live in at least equal numbers or more on habitats other than farmland have also been observed. In fact the major questions that must be raised are why should anyone consider that bird populations remain constant, given that in the past enormous fluctuations in population numbers have been recorded for birds such as the tree sparrow. As regards declines in common garden song birds, the domestic cat population (thought to be about 8 million and to have increased in number by 50% in the last 20 years) has been estimated to eliminate some 300 million young birds and small mammals, such as dormice, every year.

The feral cat population estimated at about one million individuals, will also inflict substantial (and probably even greater) damage but the extent is completely unknown. There have also been substantial increases in the population of predatory birds such as magpies and carrion crows, probably the result of increased car traffic and dead birds and mammals on roads and roadsides. All of these influences may play some part, but it is incorrect to ascribe cause and effect to just farming particularly when bird numbers are a complex system known to result from many interacting factors. Arable farming occupies about 30% of the agricultural land surface of the UK.

The tendency to blame farming for variations in bird numbers followed directly from Carson's, 1962, "Silent Spring", a title referring directly to bird populations. The assumption was that pesticides would eliminate birds by poisoning and that population numbers would drop linearly with increased pesticide use. But habitat loss is just as likely to affect bird numbers as it was in Carson's time. Direct evidence established by the Game Conservancy on the grey partridge has identified particular conventional farm practices with declines in numbers but this is the only bird that has substantive science behind it; the rest is largely speculation. However whereas there have been declines in some 13 farmland specialist species, there have been increases in 14 farmland generalist species; increases in, for example, goldfinch, chaffinch, black cap or great tit. Tree sparrows have decreased by 50% in the last 30 years whilst unsurprisingly and perhaps causally corresponding numbers of sparrow hawks have increased 50%.

The Boarded Barns experiment measured the territories of 13 bird species continuously over 9 years and found, intermittently, slightly higher levels of birds over the organic fields, although at the end of 9 years no difference in numbers was observed. In the most intensive study on birds reported by the British Trust for Ornithology, 22 farms of organic status were compared with nearby conventional farms. However no attempt was made to match managerial competence and the investigators admitted that in many cases conventional arable farms were matched with mixed organic farms.

While bird numbers on organic farms exceeded those of conventional farms in 50 of 68 individual cases, statistical significance was only established for two species. Much of the variation was attributed to field boundary effects and to hedgerow height and width, aspects of the landscape which are not unique to organic farms but are shared equally by IFM. However the benefits of organic farming to birds in terms of biodiversity and numbers were only present in one year out of three. There is a common trend for slightly higher numbers of birds to be present on organic farms during winter but these ambiguous changes are not in any way matched by observations that no-till fields gave rise to orders of magnitude higher bird numbers compared to tilled fields in winter months.

IFM regulations advise on hedgerow maintenance and field margin width identical to organic regulations. So any increase that can be ascribed to organic should be equally shared by IFM farming except where no-till is employed where it may be higher again. Deliberate manipulation of conventional farming practice at Allerton increased bird population numbers by over 40%, substantially more than the minor differences noted on organic farms as compared to conventional farms. Those measurements indicate the potential available on any farm.

### Changing to Organic Farming would Lower Costs of Agriculture

Pretty et al. estimated the externality costs of UK agriculture. However the paper omitted any attempt to compensate the costs with estimates of the benefits of (a) food security supplied by UK agriculture, (b) the improvement of life expectancy of which certainly, a portion has come from a reliable food supply. In one sense, the article is therefore a critique on only the costs of UK agriculture and particularly its conventional version and critics of conventional farming have used it as such.

One component of these externalities estimated by Pretty et al. was the supposed costs of pesticide removal from river water. However these proved instead to be the cost of constructing, improved, water purification plants to fit in with new EC water purity regulations; largely to remove nitrate and reduce eutrophication in rivers and lakes. But as has been seen above, current organic farming procedures will represent no improvement in this respect. Besides the EC regulations are considered by many excessive and unnecessarily stringent. The justification on health grounds for lowering nitrate levels in tap water is certainly questionable. The costs estimated by Pretty et al. were for 1997 only and the paper states that the expense of removing pesticides from river water is about d120 million/year. A question in the House of Commons on 3 April 2000 showed that this figure was correct for 1997 but had dropped to only d11 million in 1999 and presumably to zero thereafter. The Minister involved stated that no figure was actually available for purification running costs, only for building. And anyway water has to be purified; one cannot segregate purification requirements of any one chemical from the many others to be found in sewage.

Whether the EC regulations were in any way necessary, because they were based on incorrect assumptions, should be considered. Nitrate was commonly used as an analgesic before aspirin became readily available. The scare story surrounding nitrate and nitrite has been shown to be false. Also Knutson performed a detailed study of the costs to human health and the economy of removing pesticides from agriculture. The costs of removal were much greater than the benefits. Pretty et al. estimated that half of the externality cost of agriculture came from the production of carbon dioxide, nitrous oxide and methane and the resultant effect of global warming on the environment. Removing the pesticide error indicated above from the figures of Pretty et al. above and using the gaseous losses determined from each of the main kinds of agriculture.

Equivalent costs for organic, assuming all the UK was organic, would be d79 and likewise d29 for no till. There are about 12 million ha of farmland in the UK and the population is about 60 million. On these figures each person therefore is environmentally costed d30/year for conventional agriculture, d16 for organic and d6 if it were no-till. Those bills should be set against the average estimated food costs/person/year of d1500 for conventional, IFM and no-till and d2100 for organic. The benefits from a secure food supply which conventional agriculture offers cannot easily be costed, because they involve life and death issues.

### Organic Uses Less Overall Energy

There have been claims that organic farming requires less energy because it does not use synthetic fertilisers or pesticides. Bertillson examined the energy costs of various kinds of agriculture and pointed out that the most energy efficient/ha was a conventional farm coupled with willow plantations on part of the farm to provide generating stations with combustible fuel (biofuel). In fact his figures indicated that such farms produced more energy than they consumed. However Bertillson also provided figures for the use of energy in the production, transport and distribution of fertiliser and these figures are about 10kWh/kgN. No data is available for the mining and transport of phosphate fertiliser although it seems to be forgotten that organic farming ultimately depends on its use and other fertilisers.

Reganold et al. made the critical error of assuming that, because chicken manure was purchased from a conventional farm it involved absolutely no energy in its production. Chickens however are fed on diets of conventional corn or wheat which use minerals including nitrate for growth. Only about 10% of the N that is fed to chickens is likely to emerge in manure but every gram of that 10% N in manure costs the same in energy terms as every gram N in the original mineral fertiliser. The energy costs have not disappeared because they have passed through the gut of a bird or been purchased from someone else or described as waste. They are a cost on the worlds energy use; quite simply there is "no free lunch".

Leake measured all the draft fuel used on organic, conventional and integrated no-till agriculture over a 2-year period. Expressed as energy used/tonne of wheat yield, organic fields used 3 × as much energy as integrated no-till fields and 2.5 × as much as conventional fields. Much of the extra energy was required for weeding and of course deep ploughing. Although the organic energy requirement can be reduced by using hand weeding, getting labour to carry out this physically damaging activity is not only difficult but extremely suspect from a health point of view. The human spine is known to be easily damaged by continuous bending. One study has also pointed to the likely physical damage of pure organic farming from heavy lifting and excessive and unhealthy labour.

Uhlin estimated that fuel represented 26% of the energy use in conventional farming; fertiliser production only accounted for 14%. Putting the Bertillson figures (for fertiliser and transport) with those of Leake on fossil fuel use on farms and Uhlin estimates of energy use, the efficiency of conversion of fossil fuel use into seed energy can be estimated. From Table 2, the figures for organic in kWh/ tonne of yield of wheat are organic 200, integrated 132, conventional 140. The latter two forms of farming are on this basis more efficient in their conversion of energy into product.

There are several caveats in the figures above. Leake used 180 kgN/ha for wheat production (Table 2). Other farmers may use more or less N, dependent on manure input. Leake however states his organic yields are substantially higher than the average UK organic yield. Bertillson assumed that intensive farming only uses 100 kgN/ha. Taking account of these considerations places integrated no-till as about 2 times as energy efficient as organic farming and conventional farming about the same.

### BENEFITS TO THE USE OF GM HERBICIDE TOLERANT CROPS

All of Faulkner's claims have been established by measurement and indicate the numerous benefits of notill agriculture over organic and conventional ploughing technologies. Organic farmers need to plough not only to remove weeds, which accumulate in organic fields, but to mineralise N and possibly P and S. Thus organic farming shares all the problems that come from ploughing and comparisons with ploughed fields apply equally to them as well as conventional ploughing. Notill agriculture is most easily introduced with herbicide tolerant (HT) crops to avoid weed problems. No-till has not been greatly used in the UK. DEFRA have issued two booklets describing case histories and indicating that clay soils are best for no-till. Much of southern and midland England is certainly suitable for no-till. There have been experiments some 30 years ago in Scotland on no-till but cold soils may have been a problem in its introduction.

However Scottish LEAFUK farmers are experimenting with min-till successfully at present. Blackgrass was also a problem in early no-till but that is now controllable with glyphosate. However this would seem to be an area for further research to be undertaken. Boarded Barns also used no-till integrated farm management as part of its experiments and other experiments were performed at Stoughton. However the no-till fields at Boarded barns were only cultivated for 4 years and differences emerging in environmental measurements over a 4-year period are usually insufficient to assess a farming process that requires basically a permanent commitment to no-till on a field basis.

**Advantages of No-till Agriculture Compared to Tilled Organic and Conventional Fields**

1. Farm fossil fuel use (on a yield basis) reduced to 1/ 3rd that of a tilled field.
2. Soil erosion is reduced to 5% of a ploughed field. Soil nutrition, structure and drainage are vastly better and flooding all but eliminated. Preventing sediment losses from erosion greatly improves the local aquatic habitat. Estimates suggest that erosion causes losses of 9 billion dollars in the US. However from 1982–1997, soil erosion diminished by an estimated 30% the result of an extensive introduction of zero tillage agriculture.
3. Run-off of fertilisers and herbicides is greatly diminished and herbicides such as isoproturon when used are no longer detectable in soil drainage. Rain water preferentially drains through a no-till soil and

the longer residence of chemicals in the soil enables microbial action to break them down.

4. Free oxidised nitrogen compounds such as nitrate are reduced to 1/5$^{th}$ to 1/20$^{th}$ that in tilled soil and free phosphate to 1/6$^{th}$. Thus nitrate in waterways is reduced by this amount and far lower than any EC regulations specify.
5. Pest predators increase six-fold in number. Biodiversity has not been assessed. These pest predators hide under the vegetation on the surface and are thus largely unaffected by sprays on the canopy. They are also protected from many bird predators.
6. Large earthworm numbers increase by up to 6-fold. The latter greatly improves drainage by leaving open channels in the soil. In ploughed soils only much smaller worms predominate.
7. Bird territories and bird nests increase anywhere from 3–100-fold. The time requirements for young bird feeding are reduced five-fold. These figures contrast quite strongly with the very minor and uncertain differences observed between well-managed conventional farm and organic farms.
8. Small mammals are more abundant in no-tilled fields.
9. Soil moisture is better balanced during drought. Lack of ploughing enables the survival of capillaries that connect the surface layers with water tables at much deeper levels so during drought sufficient water rises up for the growing crop.
10. Soil carbon accumulates in the surface layer and the contribution to global warming reductions can be substantial. However one pass of the plough leads to an enormous increase (14-fold) of carbon dioxide evolution. The input of oxygen into the soil enables large-scale degradation of organic material by microbes. Robertson et al. showed that no-till has 1/3$^{rd}$

global warming potential of orgánic which in turn is about half that of large scale conventional farming in the USA. The precise similarities of US agriculture and UK agriculture are uncertain. Intriguingly these authors also reported that methane and nitrous oxide losses from soil were similar amongst all farming modes although these measurements were made on US farms and could be different in UK soils. Mineralisation to produce N for crop growth also requires input of oxygen into the soil but results in inappropriate release when crops are not present to remove free N from the soil.

11. No-till mimicks seasonal change in soil surface vegetation in both meadow and prairie.

**REFERENCES**

Bazzaz, F.A., *Plants in Changing Environments*, Cambridge University Press, Cambridge, 1996.

Dich, J., Zahm, S.H., Hanberg, A., Adami, H., "Pesticides and Cancer", *Cancer Causes Control*, 1997.

Kovalchuk, I., Kovalchuk, O., Kaick, V., Boyko, V., Filkowski, J., Heinlein, M., Hohn, B., "Pathogen-induced systemic plant signal triggers DNA rearrangements", *Nature*, 2003.

Newsome, R., "Organically grown foods", *Food Technology*, June 1990.

Simon, J., *The Ultimate Resource*, Princeton University Press, Princeton, 1996.

Sinclair, T.R., Cassman, K.G., "Green revolution still too green", *Nature*, 1999.

Smil, V., *Transforming the World: Synthesis of Ammonia and its Consequences*, MIT Press, Cambridge, MA, USA, 2000.

________________., *Feeding the World*, MIT Press, Cambridge, MA, USA, 2000.

Trimble, S.W., Crosson, P., "US soil erosion rates-myth and reality", *Science*, 2000.

# 13

# Organic Farming: Current Trends and Future Directions

The modern face of organic agriculture emerged in the late 1960s, when farmers and consumers began to recognize that the enormous amount of chemicals being used in both crop and animal production could have dire consequences for the earth and its people. Although its initial modern image was of health nuts and hippies, organic agriculture has moved into the mainstream. It is now the fastest growing food sector. From coffee in Mexico to tea in China, from shrimp in Thailand to beef in Argentina, from cotton in Uganda to olives in Italy—today's organic farmers build on the past, using modern scientific research to guide them in upgrading their traditional farming methods.

Science can now explain how and why those methods worked, opening avenues for even further improvement. New information emerges constantly from scientific research, uncovering nature's incredible secrets like pieces in a gigantic ecological puzzle—bees pass through natural foot baths in their hives that reduce mould in strawberries; bats, long considered crop pests, actually provide many services such as pollinating, dispersing seeds and controlling insects; earthworms, bacteria and fungi serve as soil "engineers" and create conditions for water and nutrients to circulate. The more farmers know about these puzzle pieces, the better their ability to manage the natural elements and benefit.

The organic movement exists amidst controversy. Doubters express concerns that without synthetic fertilizers and pesticides there will not be enough yields to feed the earth's growing population, that the market is driven by developed countries whose strict regulations cannot be met by developing-country producers, that organic agriculture's labour-intensive production methods increase drudgery. Movement supporters point out that crop yields often decline after extended chemical use, increased labour requirements of organic agriculture balance with decreased expenditures for inputs, organic products do not have to be certified to be legitimate, and that there is no way to put a monetary value on conserving the earth's biodiversity and fragile ecosystems as well as its local cultures and culinary traditions.

Absence of synthetic inputs is not the only prerequisite of organic agriculture. In fact, traditional systems that are "negligent", that carelessly mine soil nutrients or overgraze pastures, would not qualify for any organic status, even if there are no inputs present. Traditional farms do not need to be certified to be considered "organic"; when farmers follow organic management principles, thus producing food while conserving natural resources, they qualify as "non-certified organic".

The type of production methods organic farmers choose depends on more than agro-environmental conditions. It also depends on socio-economic situations, such as their labour and investment capacity and on the targeted market. Organic farmers put great efforts into study and training, seeking out developments appropriate for their farms, adapting them, and eventually going through inspection and certification to access markets, all aimed at creating a sustainable system throughout the food supply chain. The goal is for fields and animals to produce to their maximum and for farmers to protect the quality of their air, soil and streams. As the organic food system grows, so does its ability to connect people with their environments.

### FAIR AND ETHICAL ORGANIC FARMING

In the midst of a globalized economy, small farmers tend to get lost. With the advent of free trade, with wealthy countries subsidizing their farmers so their exports undercut local farmers' prices, with agricultural lands in the hands of a few powerful owners and multinational corporations controlling most of the world's food supply, with the emergence of patents for biotechnology and seeds, and with the existence of policies that favour imports over domestic production, small-scale agriculture itself is facing degradation. In their struggle against seemingly helpless poverty, small farmers out of desperation become agents of their own destruction, overexploiting natural resources because of land scarcity and lack of economic opportunities. Some abandon their land altogether and go to the cities looking for other ways to earn money, leaving behind the only way of life they understand.

The organic agriculture movement offers farmers new possibilities for helping themselves. Although considered modern, organic agriculture actually is based on traditional agriculture knowledge and culture. Over the centuries, traditional agriculture developed because small farmers experimented, innovated, adapted and worked together. They shared their seeds and new ideas about how to improve production. Today's organic agriculture has the same evolution, recognizing there is no one way to address a problem, every case has a specific response, and the concerned farmer is the one who knows best. Farming methods evolve to match local environments, responding to unique biophysical and socio-economic constraints and opportunities.

Increased trade, with its exchange of goods, capital and labour, has also meant the erosion of small farmers' knowledge and collapse of rural institutions. Too often, decisions on what to grow or how to farm have been decided by outside "experts".

Organic agriculture, with its emphasis on local resources and local ecological knowledge, brings farmers together in their communities. The market demand for organic products, as well as the opportunity to develop agri-tourism for city dwellers who appreciate a healthy, diversified rural landscape, have created new income opportunities for organic farmers.

Organic farmers' and consumers' groups work to support markets, cut out monopolies and increase farm incomes. Organic farmers, both in developed and developing nations, recognize the need to join with their neighbours, set up support networks, share what they know, pool their resources, establish themselves as an organized force and, of course, provide quality products.

Consumers increasingly demand both environmentally friendly and socially just products which means that organic certification also requires proof of fair wages, healthy working conditions and the workers' right of association.

With the organic movement, it is more important than ever for farmers to be connected. Organic farmers cannot operate in isolation because in order to maintain the integrity of their air, water and soil, they depend on their neighbours' farming practices.

The deep roots of the organic agriculture movement connect farmers, consumers and their markets, improving economic conditions and creating a vibrant rural community.

#### MORE INCOME

Modern agricultural methods have brought spectacular

increases in productivity—from fields, from animals and from people. But in spite of all the research that has gone into increasing crop yields, growing larger animals, and saving labour through chemical herbicides, pesticides, fertilizers and new technologies, and in spite of the fact that food production has increased dramatically, almost 800 million people still go hungry.

Most farmers in developing countries are poor. They either cannot afford or cannot find the necessary inputs to improve their production, and they are not connected to markets to buy or sell food. So as the large landholders who have the resources to purchase the inputs and learn to use them are able to increase their holdings and their incomes, and as international trade brings down prices even in local markets, small farmers become less and less able to compete or, in some cases, even to survive. Ironically, one solution for these small farmers has been on their farms for centuries. What is now called "organic agriculture", put forth as a modern solution to environmental problems caused by chemical-dependent farming, is actually an improved form of the farming traditionally practiced by small farmers—farming that relies on efficient use of available resources such as soil, water, local breeds and varieties of animals and crops, and a mix of modern and traditional ecological knowledge.

For countless generations, small farmers in developing countries have inherited and managed complex farming systems adapted to their local conditions. These individual farming systems have helped them survive harsh environments, meet their family's food needs, contribute to local/regional markets and still conserve their natural resources. Many of these farmers would be amazed to know that what they learned from their parents is now part of a modern, global movement.

Another irony is that many small farms that never had the advantage of technological advances could now be considered "organic", although "organic by default". An estimated 450 million farms were by-passed by Green Revolution technologies such as mechanization, irrigation, improved seeds and breeds, and synthetic fertilizers and pesticides. These include countless low-level technology systems which feature polycultures, agroforestry, integrated crop-livestock systems and productive agro-ecosystems well adapted to local conditions. The absence of these technologies does not automatically make farms organic, but makes it easier for interested farmers to upgrade their farms to organic by adding modern scientific organic practices to their traditional farming methods.

Organic agriculture stresses diversification and adaptive management which significantly decreases vulnerability to weather vagaries or other factors. In spite of a misconception that agro-ecological systems cannot increase agricultural productivity, the reality is that multicropping increases yields significantly.

A diversified organic farming system increases farm production by 20 to 60 percent as compared to a traditional low-input system and, at the same time, improves stability by improving soil and water quality and the ecological services that support agriculture. These methods may not achieve the same yields as the high input systems, but they provide a path to increase both yields and incomes. Farmers can use available resources without fear of major losses, such as those incurred by farmers who often see their expensive inputs washed away by unexpected rains.

It all adds up to a direction for improving food security that is highly promising. Farmers do not need to be certified to know that if they adopt organic agricultural systems and yields become healthier, more dependable and plentiful, their ability to feed their families, either with the food they produce or income from products they sell, will also increase. Even in areas with poor resources and no access to markets, farmers can improve their family food supply if they have access to extension or training to re-learn to farm organically, and are willing to replace the external inputs with their own labour. The result will be an improved return on their labour investment in terms of better income. The success of organic agriculture in improving food security is there for those who wish to see it. What is needed now is more studies to increase understanding of why it works and to develop methods for scaling up and sharing what has been learned.

## CREATING NUTRIENT CYCLE

Well-managed organic farms are easy to recognize. The soil looks soft and dark from the addition of compost. Often, there is mulch on the ground. Occasional weeds appear between the crop rows. Flowers and hedges are planted near or between the fields to encourage pollination and provide habitat for the helpful insects that feed on crop pests. The landscape is diversified because trees and animals are often integrated with the cropping system. In addition, in the fields, farmers and their neighbours work together.

Organic agriculture enhances people's ability to live in harmony with nature and to derive economic benefit from their land. Forty percent of the earth's land is agricultural cropland, managed forests or pasture, so it stands to reason that if agriculture does not exist in harmony with the rest of the earth's environment, the potential for widespread environmental degradation caused by loss of habitats and forests, and by pollution of air, soil and water from agricultural chemicals, looms quite large. This enormous piece of the earth's real estate both affects, and is affected by, climate change and quality of soil, water and air.

Inappropriate agricultural practices account for 28 percent of the soil erosion of the past 50 years. Scientists warn that as global warming becomes more and more of a recognizable reality, cultivation zones will shift, and pests and diseases will mutate and proliferate. Through intensification of livestock systems, nitrogen fertilization and irrigation, agriculture contributes to more than 20 percent of global anthropogenic greenhouse gas emissions.

Soil fertility is the cornerstone of organic management. Because organic farmers do not use synthetic nutrients to restore degraded soil, they must concentrate on building and maintaining soil fertility primarily through their basic farming practices. They depend on multicropping systems and crop rotations, cover crops, organic fertilizers and minimum tillage to maintain and improve soil quality. The natural fertilizers they use, such as green manure, farmyard manure, compost and plant residues, build organic content and increase the soil's capacity to circulate nutrients, air and water. As crops use soil nutrients, they can be replaced with natural rock minerals such as potassium, phosphate, calcium, magnesium and other trace elements from external sources. Organic agriculture stresses careful management to meet crop needs and avoid excess application of manure and other organic matter that could cause nitrate leaching.

In Europe, soils managed organically have 30 to 40 percent more biomass and 30 to 100 percent more microbial activity than soils managed conventionally. In some developed countries, government waterworks encourage conversion to organic agriculture to reduce the cost of purifying drinking water. In many developing countries, organically managed soils have substantially less erosion and better moisture holding capacity—an essential factor in rain-fed agriculture.

In organic agriculture, the restricted use of mineral fertilizers reduces the use of non-renewable energy (fossil fuels) and reduces the emissions of agricultural greenhouse gases (carbon dioxide, nitrous dioxide and methane). Moreover, mixed farming and soil building allow for increased biological activity by providing support for micro-organisms, earthworms, fungi and bacteria. Soils enriched with fauna and flora not only increase nutrient cycling and agricultural productivity but stabilize soils against erosion and floods, detoxify ecosystems and may even help counteract climate change by restoring "soil's capacity for carbon sequestration".

The positive impact of organic agriculture practices on air, soil, water and biodiversity offers opportunities to implement international environmental agreements such as the Convention on Climate Change (the Kyoto Protocol), Convention on Biological Diversity (Decision III/11 on the conservation and use of agricultural biological diversity) and national strategies to implement the Convention to Combat Desertification.

### WATER AND SOIL CONSERVATION

Desertification, or the degradation of drylands, is caused by overcultivation, overgrazing and deforestation, and results in soil exhaustion and soil erosion. It diminishes soil productivity, reduces food production and deprives land of its vegetative cover. It also has negative affects on areas not directly involved, because desertification can also lead to floods, soil salinization, deterioration of water quality, and silting of rivers, streams and reservoirs.

Organically managed soils are more resilient both to water stress and nutrient loss. Because of this, they have the potential to counter soil degradation. Organic farmers feed their fields with organic matter that enhances degraded soils. Micro-organisms have a good feeding base and create a stable soil structure. Water and nutrient retention capacity increases thanks to a high level of organic matter and permanent soil cover, allowing a substantial reduction in the amount of water needed for irrigation.

Although there has been little specific research on organic agriculture's potential for combating desertification, several practical trials of organic agriculture systems in arid areas have found that organic agriculture helps bring degraded lands back to fertility and create suitable microclimates in dry areas. For example, 70 ha of desert land near Cairo were converted into fertile soil, able to support livestock and bees, through organic and biodynamic agricultural methods including composting, mulching and cover cropping. In Kenya and Ghana, organic agroforestry projects were found to fight drought, control erosion and retain moisture.

Other trials studied organic agriculture's ability to counter erosion. In the tropics, even flat lands erode with the use of herbicides and the lack of soil cover. In organic agriculture, permanent soil cover is an intrinsic part the system. Locally adapted leguminous crops have been found to restore degraded soils very quickly, suppress weeds, fix nitrogen and prevent erosion. When properly set up and adapted to soil and climatic conditions, they do not compete with the main crops for nutrients or water. Even though the adoption of organic agriculture seems a viable alternative in arid areas, there is a range of constraints to adopting organic agriculture systems or even individual techniques. These include a lack of knowledge, scarce availability of organic materials, insecure land tenure, the perception of organic agriculture as being old fashioned, and the fact that extension services promote a way of farming dependent on chemical inputs.

In countries suffering desertification, the promotion of organic agriculture through education and training could be key to bringing degraded land back into production. More importantly, sound soil management practices can prevent land degradation and further agricultural encroachment in fragile areas.

### ADAPTED BREEDS AND SPECIALITY PRODUCTS

Organic agriculture requires an enormous commitment from farmers—commitment to a form of agriculture that is knowledge and labour intensive, requiring understanding of local varieties and breeds and of local environmental conditions. However, it is also economically advantageous for farmers to use these local products that are resistant to disease and extreme conditions, and thus more likely to survive, to make it to market.

Many of today's so-called modern farmers have turned to monocropping systems that require store-bought improved seeds and breeds and massive amounts of synthetic inputs and veterinary products meant to increase yields and supply markets. Tragically, many of the varieties that represent an important gene pool for resource-poor farmers are vanishing, replaced by high-yielding, chemically dependent hybrids, and more recently genetically modified organisms, developed by large seed companies.

The organic agriculture movement, with its focus on quality, nutrition, resistance to stress factors such as poor soils, harsh climates or diseases, its reliance on locally adapted plants and animals, and its goal of increasing yields, stands as an oasis in the midst of that world. Organic farmers must commit to establish a minimum level of biodiversity to compensate for the restriction on synthetic input usage.

Valuing and improving farmers' indigenous knowledge is a very important part of today's organic agriculture movement. Traditional farming, fishing and forest communities not only understand and can apply good ecological theory in decisionmaking, they understand global environmental threats and their potential impact on their livelihoods. Farmers are not trained scientists, but they are ecologically literate. Their knowledge has been acquired through on-farm experimenting, observation of natural processes, and sharing.

Throughout the centuries, farmers have developed local breeds and species of plants and animals, nurtured them and adapted them to local conditions. They saved their own seeds for planting from year to year. When their crops were inundated by pests or there were harsh weather conditions, they saved the seeds of the most resistant plants so that their resistance would be manifest in future generations. In difficult climatic conditions and in the absence of veterinary services, farmers bred local animals to survive adversity and resist diseases. The primary production strategy of organic farmers is to maintain traditional plants and animals that may yield less in the short run but will be more resilient and better able to survive in the long run.

Diversified organic farming systems have built-in insurance policies—if one crop fails, others will provide sufficient food for the family. Wild species also provide services within organic systems such as pollination, pest control and maintenance of soil fertility, all of which increase biodiversity and, in turn, support agricultural production.

Consumer demand for traditional and specialty products creates new market opportunities and ensures the economic viability of traditional products. For example, the commercial revival of organic quinoa and naturally coloured cotton from Peru, of organic chocolate processed according to an ancient Mayan recipe in Mexico, of local rice varieties in Indonesia, and of native poultry in South Africa has created connections between marginalized producers and the market. Organic agriculture's economic valorization of endangered genetic resources is key to the survival of poor rural communities in marginal areas as well as of under-utilized species and varieties. An organic market can add value to both the local area's economy and its biodiversity.

## AGRICULTURAL LAND AND WILDLIFE HABITATS

Setting up protected areas such as national parks or nature reserves is a common solution to the problem of conserving biodiversity. But a protected zone that provides a safe environment for plants and animals is just one small piece in a complex, global environmental puzzle. If that piece does not fit snugly with other pieces of the environment, then the gaps will defeat the purpose of the protected area. After all, birds and animals cannot read warning signs, seeds blowing in the wind do not stop spreading because there is a chain link fence, insects burrowing through soil never know they have entered or left a protected area, and seeping water does not carry a compass.

In other words, if protected zones border conventional agricultural lands, there is no way to stop the seepage of pollutants, there is no way to warn birds and bees not to land on fields sprayed with potentially dangerous chemicals. Moreover, many of the world's recognized ecoregions are in tropical and semi-tropical regions, populated by people living in poverty. Agriculture, as practised today, has the potential to cause serious threats in many of these regions. Forty percent of the earth's land is agricultural and another 12 percent is protected.

Organic agriculture offers a solution for connecting all of the pieces. Organic farming methods can reduce the detrimental effects of conventional farming and, in fact, can help restore natural resources and increase production. With its focus on crop rotations, natural inputs, local breeds and species, and family activities, organic is a natural link between agricultural practices and the biodiversity that needs to be protected. If farm land bordering and connecting protected areas employs organic methods, there is no reason to fear the loss of wildlife or contamination of air, water and soil. These buffer zones are critical to the success of conservation in the protected areas.

Organically farmed areas not only provide safe buffer zones between protected areas, they increase the impact of the protected area itself. Many studies document benefits in terms of protected and improved habitat for native wildlife. From micro-organisms in the soil to migrating birds, endangered species not only return, they often flourish in organic settings. The absence of synthetic pesticides and the presence of hedges and higher crop density on organic farms mean safe access to food and shelter for wildlife species.

In addition, organic farms can operate safely within certain types of protected areas. There are numerous examples of successful organic agriculture inside protected areas such as Muraviovka Park, Russia (wetlands), Yaoluoping National Resource Reserve, China (mountains), Monte Azules Biosphere Reserve in Mexico (tropical forests), and the Meso-American Biological Corridor that stretches across seven countries.

Most protected areas traditionally belonged to local villagers, so organic agriculture allows local people to maintain some management control over land they consider theirs. Organic farmers protect land through their farming practices, reap its benefits for themselves and, at the same time, they conserve and improve their natural environment which benefits the entire planet.

## GUARANTEEING THE VALIDITY OF ORGANIC SYSTEM

With organic products, the label tells a big story. Informed consumers know that the label will tell them everything from the name and type of organization that certified the product's "organic" validity to the type of standards that the product met throughout its growth and processing. These standards make organic agriculture the food industry's most well defined food supply system.

Certification bodies ascertain that products meet the standards set by either private or public institutions. In countries that have organic legislation, certification bodies are approved and supervised by the government. Some also are accredited by the International Organic Accreditation Service (IOAS), an independent non-governmental organization that authorizes certifiers who follow the voluntary standards of the International Federation of Organic Agriculture Movements (IFOAM).

Administered jointly by the Food and Agriculture Organization of the United Nations (FAO) and World Health Organization (WHO), the Codex Alimentarius Commission develops and adopts international food standards, guidelines and related texts that serve as the basis for many national standards. It governs such crucial issues as food labelling, residues of pesticides and veterinary drugs, food hygiene, food import and export inspection and certification systems. In 1999, Codex adopted guidelines on the Production, processing, labelling and marketing of organically produced foods that provide a description of the "organic" system and the "organic" labelling claim in order to ensure fair trade and to facilitate the development of the organic sector.

The private sector equivalent of the Codex Alimentarius guidelines is the International basic standards for organic production and processing, created by IFOAM. Both the Codex and IFOAM guidelines cover the production of plants, livestock and bees; the handling, storage, processing, packaging and transportation of products; and a list of substances permitted in the production and processing of organic foods. IFOAM also has provisions for fibres, aquaculture and non-wood forest products.

Governments can use these guidelines to develop their own national organic agriculture programmes, tailored to their specific needs. Most national standards, such as those of Argentina, European Union countries, India, Japan, Tunisia and the United States of America, set forth regulations which are legally binding. In the major organic markets (i.e. Japan, European Union countries and the United States of America), certifiers are approved and supervised by the national governments.

The private sector, and certification bodies in particular, have so far been the main agents in guaranteeing the validity of the organic system, ascertaining that farms are inspected according to agreed upon organic standards before products are certified "organic" for labelling. It is a system that aims at ensuring credibility and building consumer trust.

### RULES OF ORGANIC AGRICULTURE PRODUCTION

The rapid growth of organic agriculture production and trade has been accompanied by an increase in governmental regulations. Regulations establish rules for organic farmers and processors through standards, give credibility to certification bodies through approval and supervision, protect consumers against mislabelling and fraud through conformity and surveillance, and eventually can facilitate trade through equivalence agreements with export countries. Regulations usually include provisions that further promote organic agriculture such as support to research, production and marketing.

At the regional level, the European Union (EU) legislation on organic production, Regulation 2092/91, was adopted in 1991 and is amended yearly. The Regulation sets mandatory standards for organic agriculture with rules on production, labelling and inspection. The aim is to ensure fair competition among organic producers and to gain credibility among consumers. EU Member Nations support the Regulation with domestic legislation, creating their own inspection and certification schemes. Member Nations are required to meet the Regulation's minimum requirements, but they may enact stricter legislation. In the last decade, many developing countries and countries in transition used Regulation 2092/91 as a model for designing their own national legislation, to assure they will meet EU import requirements and thus enhance their export opportunities.

Along with the European Union, Japan and the United States of America are the main organic markets. Japan began implementing its organic agriculture regulations in

2001, the United States of America in 2002. While similar in many respects, there are also several dissimilarities, both in coverage and substance. This means that exporting countries must be aware of the variations in specific requirements in order to access the different markets. Of course, this need to comply with several regulatory systems increases costs at the production, certification and accreditation levels.

According to a survey conducted by IFOAM in 2002, about 40 nations have completed the process of enacting their organic legislation, including detailed standards and institutional arrangements. Another 15 or so nations are in the process of developing organic legislation. Government regulations have grown steadily, but often in an uncoordinated manner. The resulting variation in production, inspection and certification systems hinders farmers' access to regulated organic markets and the proliferation of organic labels can be confusing to consumers.

Compliance with foreign organic standards is not always suitable to agroecological conditions of exporting countries. Inspection and certification costs may be out of reach for small producers. Alternative systems, more relevant to small farms and local marketing, are being developed by the private sector. Many countries have developed Internal Control Systems and in the United States of America, a Certified Naturally Grown labelling programme has been created.

Standards, inspection, certification and accreditation (or the organic guarantee system) ensure credibility. There is need for a more harmonized approach to regulate the sector, to ensure that organic rules and standards are designed and applied consistently and fairly, taking into account all stakeholders' interests. Flexible requirements to establish equivalency among regulatory systems is the main challenge ahead for the organic agriculture community. FAO, together with the United Nations Conference on Trade and Development (UNCTAD) and IFOAM are looking into ways to harmonize the organic guarantee system. FAOLEX, FAO's database on food and agriculture legislation, contains regulations and standards dealing with organic agriculture in some 30 countries and the EU, which are accessible online through the organic agriculture website.

### FARMERS AND CONSUMERS

Farmers' reasons for adopting organic agricultural practices are often the same reasons consumers have for buying organic products. Both farmers and their customers know that organic farming methods that avoid synthetic pesticides and fertilizers and minimize pollution of soil, air and water, are good for the environment and produce food that is perceived as healthier. But beyond the benefit to the earth and its inhabitants, farmers also must consider another benefit before entering the organic world—the financial benefit.

Organic is the fastest growing food sector, in both land use and market size. Yet, official statistics are non-existent and global assessments are scarce. National statistics gatherers rarely recognize organic production as worthy of being reported separately. Organically produced commodities are not identified as such in international trade. Organic products consumed or marketed locally often are not submitted to certification. In addition, when there is no organic market, certified organic products sell as conventional produce and are not reported in organic data.

Global retail sales for certified organic products are estimated to reach US$23-25 billion in 2003, but no specific numbers exist for organic production, consumption, trade or prices. Organic agriculture's share of food and beverage sales may be small, 1 to 2 percent, but its move from specialty food stores onto supermarket shelves, where up to 15 percent of some products in specific markets are organic, indicates the movement has mainstreamed. This increased exposure also forewarns the potential for multinational food companies and retailers to become major players in organic food supply, in terms of contracting production, supermarket offers and international trade. If they are to compete, small producers in developing countries need access to the required know-how for production, post-harvest handling and marketing, as well as to affordable certification systems.

Organic products receive better prices in the market. But going organic to take advantage of the added value requires a long term commitment. It takes time to cleanse land of chemicals used in modern farming practices and to learn techniques of good organic management. Evolving issues such as farm subsidy policy, expensive certification processes and demand-supply balance affect prices. The intricate cropping system is knowledge and labour intensive and thus more expensive in the start-up phase. Consumer markets are growing but reactive, driven by food safety concerns. The higher retail prices of organic products are affected by economic swings. In addition, organic agriculture is most successful when it relies on local species, varieties and breeds that are more resistant to disease and grow better in local environments, but are not well known to consumers. On the other hand, organic farmers do not have to pay for synthetic inputs. In addition, they often establish producer-consumer groups to provide direct food marketing through such activities as farmers' markets or home deliveries to subscribed customers, which increases profits.

A high volume of marketed organic produce is channeled to general food shops, including supermarkets, by wholesalers and distributors. The increase of market share of organic products is greatly dependent on the involvement of general food retailers in the organic food market because it lowers costs and thus expands the consumer base.

In order for the organic sector to thrive, it needs more than growing markets. It needs governmental and intergovernmental policies to harmonize standards; provide development support through research and advisory systems, education and consumer information; regulate use of synthetic inputs; and implement global environmental treaties. In addition, public and private investments in organic agriculture research are needed to balance the research and extension programmes that promote intensive use of synthetic inputs.

## SATISFYING CONSUMERS

At first glance, the market for organic products in developed countries appears to offer great export opportunity to developing countries, but in reality, the situation is quite complex. In order to be successful, an exporting country needs to maintain a competitive price while meeting organic and health standards and providing the same quality as conventional products.

In most developed countries, the demand for organic fruit and vegetables has been growing at 20 to 25 percent a year for the past decade and further market expansion is forecast, although at a slower rate. The share of organic products in the fruit and vegetable market of these countries is estimated between 3 and 7 percent. Domestic organic production in developed countries is expected to rise, but it is unlikely to meet demand for many products, thus providing export opportunities. However, consumers traditionally prefer to buy locally produced organic fruit and vegetables, and they tend to distrust the certification procedures of organic products imported from developing countries.

Farmers considering conversion to certified organic production should bear in mind the need for adopting new production and management methods. One attraction of organic agriculture is the premium price paid for the products, but the price difference between organic and conventional products is expected to decrease. Therefore, both producers and exporters should assess carefully the potential of their product in the targeted market and identify competing suppliers. There are many steps that must be taken before organic fresh fruit and vegetables can be exported successfully:

— Learn organic farming methods: organic agriculture requires knowledge of the alternative practices that make organic farms successful without using synthetic inputs.

— Obtain certification: certification can be obtained locally for countries that have national organic standards and control systems that are equivalent to the organic rules of the country to which exports will be sent.For countries that do not meet these conditions, farmers and producers need to work through an accredited organic certification organization or an importer specialized in organic produce.

— Set up production infrastructure: for fresh produce to arrive in good condition in the country of destination, good post-harvest handling, such as cold storage, and good infrastructure and logistics, including harbours and airports, are necessary in the exporting country.

— Establish trade relations: exporters must have good and reliable relations with an importer, trader or wholesaler in the target market in order to be successful. The importer will be aware of current developments in markets and regulations.

## RESEARCH AND DEVELOPMENT

The modern organic agriculture movement evolved in

developed countries, mostly in temperate regions. Now, with growing interest in organic cultivation as a management method for agricultural production in tropical and subtropical countries, greater attention needs to be given to developing standards and guidelines for organic agriculture applicable to tropical products and ecosystems.

Motivation for converting to organic management can include market demand, lack of access to inputs, environmental concerns, social concerns or personal philosophy. In addition, farmers' production targets need to be considered in terms of both quantity and quality, weighing the costs of production against the expected returns. Clear and objective information about both the potential and the constraints of organic agriculture is essential for any farmer contemplating conversion.

Export marketing opportunities for developing countries include those organic horticultural products that are not produced domestically in temperate countries, such as spices and tropical fruits and vegetables.

Tropical and subtropical fruit production takes place in settings ranging from traditional home gardens to large plantations, making it difficult to prescribe organic husbandry practices that are suitable for all. Nonetheless, any fruit growers contemplating organic conversion need to have a clear understanding of all aspects of organic production and marketing before taking investment decisions.

A challenge for organic fruit producers is the maintenance of economic feasibility while complying with organic standards. This means dealing with crop-specific problems without using substances prohibited by organic standards. Alternative approaches are needed, for instance, to control ripening of pineapples without chemicals, or to control banana or citrus pests and diseases. Furthermore, conversion to organic agriculture requires innovative solutions that guarantee purity of seeds and planting material and provide assurance against the risk of pest and disease spread. Some of these problems might have partial local solutions while others require specific research to identify or generate the appropriate permitted management practices.

Organic horticulture makes optimum use of crop residues, green manure and biological nitrogen fixation, but this does not replace all nutrients removed by crops. Nitrogen supply needs to be supplemented by the provision of other essential plant nutrients that usually cannot be replenished sufficiently by organic matter in acid tropical soils. Considering the nutrient demand of perennial fruit crops over time, it is important to have adequate methods to maintain soil fertility, especially levels of phosphorus, potassium and micronutrients, in such a way that yields can be sustained. There is a strong need to set up networks for sharing empirical experience and scientific results of various organic methods of fertilization, in order to get a more complete picture of long-term successes that can be shared with organic farmers.

## PEST MANAGEMENT PRACTICES

Pest control in organic agriculture begins by making sensible choices, such as growing crops that are naturally resistant to diseases and pests, or choosing sowing times that prevent pest and disease outbreaks. Careful management in both time and space of planting not only prevents pests but also increases populations of natural predators that can contribute to control of insects, diseases and weeds.

Organic agriculture requires informed decision-making about a range of pest management practices which are being developed. This decision-making and field management process, known as Organic Pest Management (OPM), is an outgrowth of Integrated Pest Management (IPM). Although the latter allows usage of synthetic inputs, both systems emphasize prevention through such methods as:

For curative practices, OPM limits itself to those products approved by certification bodies. But even among these products, there are differences in costs, applicability and access to be considered by farmers. There has been a great deal of research into OPM, and farmers' networks and organic agriculture associations encourage exchanges of experiences. Products are approved based on their low risk for environmental and health impact as well as effectiveness in controlling target pests. It is important to point out that some products allowed for OPM may be hazardous to applicators. For example, red chili mixtures or garlic oil which are used as pest repellants can be irritating to the skin. Applicators should have proper protective gear and application equipment.

At this point, most organic products, such as larvae of pest predators, are imported by developing countries. There is a need to promote local production to reduce import dependency. Wider accessibility and low tariffs should be encouraged. Farmer education in OPM can be improved with better written materials, better field courses and, of course, exchange visits to meet with and learn from successful farmers.

IPM programmes use "farmer field school" methods to teach ecological approaches to pest management. Similar types of group study are applicable to OPM, especially when combined with testing new methodologies, crops and marketing. Study groups and mutual support groups could also be used for peer review of practices within a specific organic labeling scheme. For example, California Certified Organic Farmers uses peer review to ensure that members using its organization label adhere to its more strict standards and develop mutual support systems. Such practices already are followed by several organic associations but could be more widely adopted.

## MAINTAINING ORGANIC CLAIM

An organic label on a product represents more than proof of organic production. It shows that the entire food chain—from the farm to the table—adheres to organic standards. For example, processed food may lose its organic claim if it contains ingredients of non-organic origin or if the produce is transported in containers which have chemical residues.

The steady increase in market demand for organic fruits and vegetables in recent years has presented an export opportunity for farmers in developing countries who practice low-input agriculture. However, they must overcome significant hurdles in order to benefit from this opportunity. Apart from regulations that apply to the production of organic produce, certain post-harvest activities need to be modified to comply with organic regulations. Many post-production operations for organic produce are identical to non-organic production. For example, there are few specific requirements for harvesting organic produce.

Although some root, tuber and bulb crops require a curing period at ambient or elevated temperature to promote wound healing and ensure optimum storage life, there are no specific requirements for curing, storing or transporting organic produce. Most markets require strict attention to the size, grade, quality and maturity of produce, whether it is organic or not. Fruit and vegetables must be cleaned and graded to comply with these regulations. Although all types of packaging are authorized, there is an expectation that careful thought will go into choosing packaging with regard to its environmental impact. Degradable packaging material is increasingly requested by conscientious consumers.

All harvested fruits and vegetables should be placed as soon as possible in a storage area that is kept at the appropriate temperature. However, organic products need to be stored and transported with proper identification and physically separated from non-organic products. As for processing, freezing is the only processing method that keeps produce in a state similar to the fresh crop. Organic foods can also be processed by drying; with the use of approved processing aids such as: ascorbic acid, citric acid, tartaric acid and salt; blanching with high temperatures to destroy micro-organisms; pasteurizing to destroy micro-organisms that could contaminate the product after blanching; and with heat treatments that conserve products by destroying or inactivating enzymes and killing micro-organisms.

Juice is an ideal product for the organic market. It offers a simple and natural way to process fresh produce, preserves the majority of vitamins and minerals, and largely resolves the problem of storage. Canned produce must be prepared in a way that retains, as closely as possible, the characteristics of fresh produce. Other forms of processing include conservation with sugar which is principally used for fruit jams and purees, and by fermentation which is a chemical change caused by enzymes, bacteria or micro-organisms. There are very few approved organic post-harvest treatments for pests and diseases. Hot-water (45-55 °C) immersion, steam and forced hot-air treatments are sometimes used as organic control methods after harvest..

## SUPPORTING HEALTHY PASTURES FOR ENVIRONMENT AND LIVESTOCK

Increasing human and livestock populations, as well as increased cropland and intensification of cropping systems, affect the sustainability of natural grassland ecosystems. As grassland ecosystems are usually in fragile environments unsuited to intensive exploitation, farmers, herders, ecologists, policy-makers and economists need to look for a way to combine grassland production and protection.

Many answers can be found in the organic agriculture system. Organic production methods and certification standards applied to grasslands contribute to the maintenance of the ecosystem, add value to grasslands products, assure income generation to their users, and promote animal welfare. Yet organically certified and managed grasslands account for only a small fraction of the grasslands worldwide.

Grasslands, which are immensely different in their climatic needs, physical conditions, cultural and historical roots and economic situations, contain 24 percent of the world's vegetation and cover about 4.6 billion ha (42.8 percent of the earth's surface) of which 9 million ha are organically certified. Grasslands are of prime importance in global carbon budget, maintenance of flora and fauna biodiversity, control of desertification, maintenance of soil fertility, and the living conditions of the population.

Grasslands can be natural, made up of indigenous species and managed with few external inputs. Or, they can be temporary, used in rotation, alternating between cropland and grassland and intensively managed. Natural grasslands produce with very few or no external inputs, maintain soil fertility and ground cover, and contribute to the maintenance of biodiversity and ecosystem functions. Most natural grasslands are relegated to the more marginal lands of the South American pampas, the Eastern European and North African steppes, high mountain areas, the Sahelian, Soudano and savannah zones of Africa, and North American prairies.

In terms of meeting organic objectives, temporary grasslands offer a number of advantages including easier maintenance of high clover content, easy exploitation of residual fertility by arable crops, minimized weed development and better possibilities to provide regular, clean worm-free grazing. However, establishment costs are high.

Organic farms in regions with climate, soil type and topography well suited to cropping are likely to have a mixed crop/grassland system, while permanent grasslands are more frequent in fragile environments and low potential areas. Because organic standards restrict crop and livestock inputs, farmers need to employ preventative approaches to crop and livestock husbandry such as crop nutrient supply, weed control and animal health if they are to build and maintain a viable organic system.

Well-managed grasslands used for extensive grazing, such as those in Australia and Argentina, offer opportunities to add value to livestock products on the market through organic certification.

## REFERENCES

Bertillson, G., "Environmental consequences of different farming systems using good agricultural practices", *Proceedings of the Fertiliser Society*, December 1992. Cambridge, 1992.

Goklany, I.M., Trewavas, A.J., "How technology can reduce our impact on the earth", *Nature*, 2003.

Hansen, J.W., Jones, J.W., "A systems framework for characterising farm sustainability", *Agriculture System*, 1996.

Norwood, C.A., "Water use and yield of dryland row crops as affected by tillage", *Weeds Today*, 1999.

Orr, J.B., *Food, Health and Income*, MacMillan, London, 1936.

# Bibliography


Altieri. M., "The green revolution revisited: New needs, new strategies", Ceres (154), 1995.

Arunachalam V., "The science behind tradition", *Current Science*, 2001.

Bazzaz, F.A., *Plants in Changing Environments*, Cambridge University Press, Cambridge, 1996.

Bedggood, W., *Weed Management – Tools for an Integrated Approach*, NRE, Horsham, 2000.

Berndes G, Hoogwijk M, van den Broek R., "The Contribution of Biomass in the Future Global Energy Supply", *A Review of 17 Studies*, Biomass and Bioenergy, 2003.

Berndes G., "Bioenergy and Water—the implications of large-scale bioenergy production for water use and supply", *Global Environmental Change*, 2002.

Bertillson, G., "Environmental consequences of different farming systems using good agricultural practices", *Proceedings of the Fertiliser Society*, December 1992. Cambridge, 1992.

Buckerfield, J. and Webster, K., "Vineyard trials show value of mulches - organic matter for *management of young vines*, Australian Grapegrower and Winemaker #441, October 2000.

Burnett, V., Enshaw, T. and Sutherland, S., *Non-chemical Options for Integrated Weed Management in Grain Production*, Victorian Department of Primary Industries, 2004.

Byczynski, Lynn, "Growing great transplants", *Growing for Market*, February 1993.

Chase, C.A., *Soilborne Plant Pathogen and Pest Management with Soil Solarization*, University of Florida, Institute of Food & Agricultural Sciences, 2002.

Cousens, R. and Mortimer, M., *Dynamics of weed populations*, Cambridge University Press, Cambridge, UK, 1995.

Delucchi M.A., "Life Cycle Analysis of Biofuels", *Draft Manuscript*, Davis, CA: ITS, University of California at Davis, 2006.

Den Nijs, A.P.M. and E. Lammerts van Bueren, "Towards breeding for organic agriculture: cross pollination required between conventional breeders and organic farmers", Prophyta June 1999.

Dich, J., Zahm, S.H., Hanberg, A., Adami, H., "Pesticides and Cancer", *Cancer Causes Control*, 1997.

Dikshit, P. K., Report on 'Organic Farming' prepared by a Technical Committee constituted by Ministry of Agriculture, Govt. of India. (Convenor: P. K. Dikshit), 1993.

Douglas, C., Gerard D'Souza and T. Phipps, 'Alternative agriculture adoption: Effect of ground water contam mination and other factors', R.M publications No 91, 1991.

El-Hage Scialabba, N.E. & Hattam, C., "Organic agriculture, environment and food security", Environment and Natural Resources Service Development Department, The Food and Agriculture Organization of the United Nations, Rome, 2002.

Eyhorn F., "Organic Farming for Sustinable Livelihoods in Developing Countries? The Case of Cotton in India", vdf Hochschulverlag, ETH Zrich, Switzerland, 2007.

Flessa, H., Ruser, R., Dörsch, P., Kamp, T., Jimenez, M.A., Munch, J.C., Beese, F., "Integrated evaluation of greenhouse gas emissions ($CO_2$, $CH_4$, $N_2O$) from two farming systems in southern Germany", *Agriculture, Ecosystems and Environment*, 2002.

Fließbach, A. & Mäder, P., "Carbon source utilization by microbial communities in soils under organic and conventional farming practice", In: *Microbial Communities — Functional versus Structural Approaches*, Insam, H., Rangger, A. (eds.). pp 109-120. Springer, Berlin, 1997.

Forster, C.; Green, K.; Bleda, M.; Dewick, P.; Evans, B.; Flynn, A. & Mylan, J., "Environmental Impacts of Food Production and Consumption", *A report to the Department for Environment, Food and Rural Affairs*, Manchester Business School. DEFRA, London, 2006.

Fukuoka, M., "The natural way of farming: The theory and practice of green philosophy", Bookventure Publishers, Madras, 1997.

Goklany, I.M., Trewavas, A.J., "How technology can reduce our impact on the earth", *Nature*, 2003.

Hansen, J.W., Jones, J.W., "A systems framework for characterising farm sustainability", *Agriculture System*, 1996.

Hatfield J.L.; Sauer, T.J. & Krueger, J.H., "Managing soils to achieve greater water use efficiency: a review", *Agronomy Journal*, 2001.

Hole, D.G., A.J. Perkins, J.D. Wilson, I.H. Alexander, P.V. Grice and A.D. Evans, "Does organic farming benefit biodiversity?", *Biological Conservation*, 2005.

Holland, J.M., "The environmental consequences of adopting conservation tillage in Europe: reviewing the evidence", *Agriculture, Ecosystems and Environment*, 2004.

International Federation of Organic Agriculture Movements, IFOAM, Basic Standards for Organic Production and Processing, IFOAM Publications, Tholey-Tholey, Germany, 1998.

Kortbech-Olesen, R., "Export opportunities of organic food from developing countries", In *World Organics*, Agra Europa (London) Ltd, London, UK, 9–10 May 2000.

Kotschi, J., Müller-Sämann, K., "The Role of Organic Agriculture in Mitigating Climate Change", *International Federation of Organic Agriculture Movements (IFOAM)*, Bonn, 2004.

Kovalchuk, I., Kovalchuk, O., Kaick, V., Boyko, V., Filkowski, J., Heinlein, M., Hohn, B., "Pathogen-induced systemic plant signal triggers DNA rearrangements", *Nature*, 2003.

Lampkin, L., *Organic Farming*, Farming Press, UK, 1990.

Lampkin, N & Measures, M., *Organic Farm Management Handbook*, University of Wales, Aberystwyth, 1999.

Mäder, P.; Fließbach, A.; Dubois, D.; Gunst, L.; Fried, P. & Niggli, U., "Soil fertility and biodiversity in organic agriculture", *Science*, 2002.

Mohan Singh, "Organic farming prospects in Indian agriculture", *Souvenir of 68th Annual Convention of Indian Society Soil Science*, CSAU&T, Kanpur, 4–8 November 2003.

Natarajan, M & Willey, R.W., "Sorghum-pigeonpea intercropping and the effects of plant population density" *Journal of Agricultural Science*, 1990.

Newsome, R., "Organically grown foods", *Food Technology*, June 1990.

Norwood, C.A., "Water use and yield of dryland row crops as affected by tillage", *Weeds Today*, 1999.

Orr, J.B., *Food, Health and Income*, MacMillan, London, 1936.

Padel, S. and Lampkin, N. H., *The Economics of Organic Farming: An International Perspective*, CAB International, Wallingford, UK, 1994.

Parnes, R., "Fertile Soil", *A growers guide to organic and inorganic fertilizers*, AgAcces, Davis, CA, 1990.

Paul V. Nelson, *Greenhouse Operation and Management* (5th edition), Prentice Hall, Inc., Upper Saddle River, 1998..

Pimentel, D.; Hepperly, P.; Hanson, J.; Douds, D. & Seidel, R., "Environmental, energetic, and economic comparisons of organic and conventional farming systems", *BioScience*, 2005.

Reganold, J., Elliott, L. and Unger, Y., "Long-term effects of organic and conventional farming on soil erosion", *Nature*, 1987.

Richardson J., "Sustainable production systems for bioenergy: Impacts on forest resources and utilization of wood for energy", *Biomass and Bioenergy*, 2006.

Rissler, J. and Mellon, M., *Perils Amidst the Promise: Ecological Risks of Transgenic Crops in a Global Market*, Cambridge, MA: Union of Concerned Scientists, 1993.

Robert Aldrich and John Bartok, *Greenhouse Engineering*, Northeast Regional Agricultural Engineering Service, 1989.

Roush, R.T., "Can We Slow Adaptation by Pests to Insect Transgenic Crops?", in G.J. Persley (ed.) *Biotechnology and Integrated Pest Management*, Oxon, UK: CAB International, 1996.

Scott D.A, Dean T.J., "Energy Trade Offs Between Intensive Biomass Utilization Site Productivity Loss and Ameliorative Treatments in Loblolly Pine Plantations", *Biomass and Bioenergy*, 2006.

Simon, J., *The Ultimate Resource*, Princeton University Press, Princeton, 1996.

Sinclair, T.R., Cassman, K.G., "Green revolution still too green", *Nature*, 1999.

Sindel, Brian M. Ed., *Australian Weed Management Systems*, R.G. and F.J. Richardson, Melbourne, Australia, 2000.

Smil, V., *Transforming the World: Synthesis of Ammonia and its Consequences*, MIT Press, Cambridge, MA, USA, 2000.

Tabashnik, B.E., N. Finson and M.W. Johnson, "Managing Resistance to *Bacillus thuringiensis:* Lessons from the Diamondback Moth (Lepidoptera: Plutellidae)", *Journal Economic Entomology*, 1991.

Tenni, A. and Faithfull, I., "How to manage weeds", *NRE Landcare Notes*, 1998.

Trimble, S.W., Crosson, P., "US soil erosion rates-myth and reality", *Science*, 2000.

Wolf, K., "Gene Transfer Between Organelles and the Nucleus in Lower Eukaryotes", in E.R. Schmidt and T. Hankeln (eds.) *Transgenic Organisms and Biosafety*, Berlin and New York: Springer, 1996.

Wolfe, M. and E. Lammerts van Bueren, "A question of breeding", *Organic Farming*, 1999.

# Index